STATDISK
Student Laboratory Manual and Workbook

Mario F. Triola
Dutchess Community College
with contributions by Justine C. Baker, Peirce College, PA

to accompany

The Triola Statistics Series:

Elementary Statistics, Tenth Edition

Elementary Statistics Using Excel, Third Edition

Essentials of Statistics, Third Edition

Elementary Statistics Using the Graphing Calculator, Second Edition

Mario F. Triola
Dutchess Community College

Boston San Francisco New York
London Toronto Sydney Tokyo Singapore Madrid
Mexico City Munich Paris Cape Town Hong Kong Montreal

Reproduced by Pearson Addison-Wesley from electronic files supplied by the author.

Publishing as Pearson Addison-Wesley, 75 Arlington Street, Boston, MA 02116.

ISBN 0-321-36912-2

4 5 6 BB 08

Preface

This *STATDISK Student Laboratory Manual and Workbook,* 10th edition, and STATDISK Version 10.1 and later are supplements to the Triola statistics series of textbooks, including *Elementary Statistics, Essentials of Statistics, Elementary Statistics Using Excel, and Elementary Statistics Using the Graphing Calculator*. STATDISK is also a supplement to *Biostatistics for the Biological and Health Sciences* by Marc M. Triola, M.D. and Mario F. Triola.

The STATDISK software is designed for IBM PC and compatible computers using Windows 98, NT, 2000, ME, or XP, and it is also designed for Macintosh OS X (10.1 or later for OS X version). The processor should be a Pentium 150 MHz or higher (233 MHz for Windows XP) or PowerPC Macintosh. In addition to the RAM required by your computer's operating system, STATDISK requires 32MB of RAM. You should also have 20MB of available hard-disk space, a CD-ROM Drive, and a Printer.

Version 9 of STATDISK has been completely recoded to improve compatibility with Windows and Macintosh systems and to include dramatic improvements over earlier versions. For updates and the latest version of STATDISK, please check the Web site at **http://www.aw-bc.com/triola.**

This manual/workbook generally refers to the Windows version of the software, and there are some differences between this version and the Macintosh version. For example, you can exit the Windows version by clicking on File, then Exit; you can exit the Macintosh version by clicking on File, then Quit.

Here are major objectives of this manual/workbook and the STATDISK software:

- Describe how STATDISK can be used for the methods of statistics presented in the textbook. Specific and detailed procedures for using STATDISK are included along with examples of STATDISK screen displays.

- Incorporate an important component of computer usage without using valuable class time required for concepts of statistics.

- Replace tedious calculations or manual construction of graphs with computer results.

- Apply alternative methods, such as simulations, that are made possible with computer usage.

- Include topics, such as analysis of variance and multiple regression, that require calculations so complex that they realistically cannot be done without computer software.

It should be emphasized that this manual/workbook is designed to be a supplement to the Triola series of statistics textbooks; it is not designed to be a self-contained statistics textbook. It is assumed throughout this manual/workbook that the theory, assumptions, and procedures of statistics are described in the textbook that is used.

Chapter 1 of this supplement describes some important basics for using STATDISK. Chapters 2 through 14 in this manual/workbook correspond to Chapters 2 through 14 in *Elementary Statistics*, 10th edition. However, individual chapter *sections* in this manual/workbook generally do *not* match the sections in the textbook. Each chapter includes a description of the STATDISK procedures relevant to the corresponding chapter in the textbook. This cross-referencing makes it easy to use this supplement with the textbook. STATDISK also includes the data sets found in Appendix B of the textbook. Chapters include a beginning section in which examples are illustrated with STATDISK. It would be helpful to follow the steps shown in these sections so the basic procedures will become familiar. You can compare your own computer display to the display given in this supplement and then verify that your procedure works correctly. You can then proceed to conduct the experiments that follow.

If you experience technical problems with STATDISK, click on the main menu item of **Help**, then select the menu item of **Report a** bug. You can also call (800) 677-6337 between 8:00 a.m. and 5:00 p.m. CST, Monday through Friday. You can also send an e-mail to product.support@pearsoned.com.

I thank Justine Baker of Peirce College for contributing *Activities with STATDISK* that are found at the end of Chapters 3, 8, 10, and 13.

I thank Bill Flynn for his great work in developing the original STATDISK algorithms. I also thank Russell F. Loane and Timothy C. Armstrong of Password, Inc for their outstanding work on a previous version. The following beta testers were extremely helpful with the new version of STATDISK: Justine Baker, Robert Jackson, Caren McClure, Sr. Eileen Murphy, John Reeder, Carolyn Renier, Cheryl Slayden, Victor Strano, Gary Turner, and Henry Feldman. For this new version of STATDISK, I am very thankful to Marc Triola, MD, who has the talent and patience to completely rewrite thousands of lines of programming code. It is wonderful working with such competent and skilled professionals. Their dedication and talent are very apparent in this new version of STATDISK. Finally, I thank the Addison-Wesley staff for their enthusiastic support in this project. It is a genuine pleasure working with a publishing company committed to providing a product with the highest quality. I also thank the many instructors and students who took the time to provide many valuable suggestions.

M.F.T.
January, 2006

Contents

1

STATDISK Fundamentals

STATDISK is designed so that it uses many of the same features common to a wide variety of software applications, so tools introduced in this chapter have a universal usefulness that extends beyond STATDISK and statistics. For example, the **Copy** and **Paste** features of STATDISK are commonly included with many software programs. As you learn how to use such features, you acquire or reinforce important and general computer skills that will help you with other applications.

1-1 Installing and Updating STATDISK

First, check the web site www.aw.com/triola for updated versions of STATDISK. If the web site contains a later version than the one provided on the CD-ROM included with your textbook, download the later version. Otherwise, install STATDISK from the CD-ROM. In some cases, your instructor will have STATDISK installed on your college network. If the network is used, your instructor will provide access instructions. If a network is not used, you can install your own copy on your own computer.

IBM PC compatible computers with Windows: Put the CD-ROM in its disk drive, click on Start, then Run, and enter the drive containing the CD-ROM followed by :/Software/Statdisk/Setup.exe. For example, if your CD-ROM is in drive D, click on **Start**, then **Run**, then enter **D:/Software/Statdisk/Setup.exe**.

Macintosh computers: Copy the STATDISK program from the CD-ROM to the computer's hard drive by clicking and dragging the program icon. Start the program by double-clicking on the STATDISK icon. The CD-ROM location is Software/Statdisk.

1-2 Entering Data

Your first use of STATDISK is likely to occur with topics from Chapter 2 of your Triola statistics textbook, and one of your first objectives is likely to be entering a set of sample data. (The sample data found in Appendix B of the Triola textbook are already stored with STATDISK, and they can be retrieved as described in Section 1-4 of this manual/workbook. It is not necessary to manually enter those data sets that are already stored with STATDISK.) To manually enter a set of sample data, use the following procedure.

STATDISK Procedure for Entering Data

1. After opening STATDISK, you should see the data window shown on the following page.

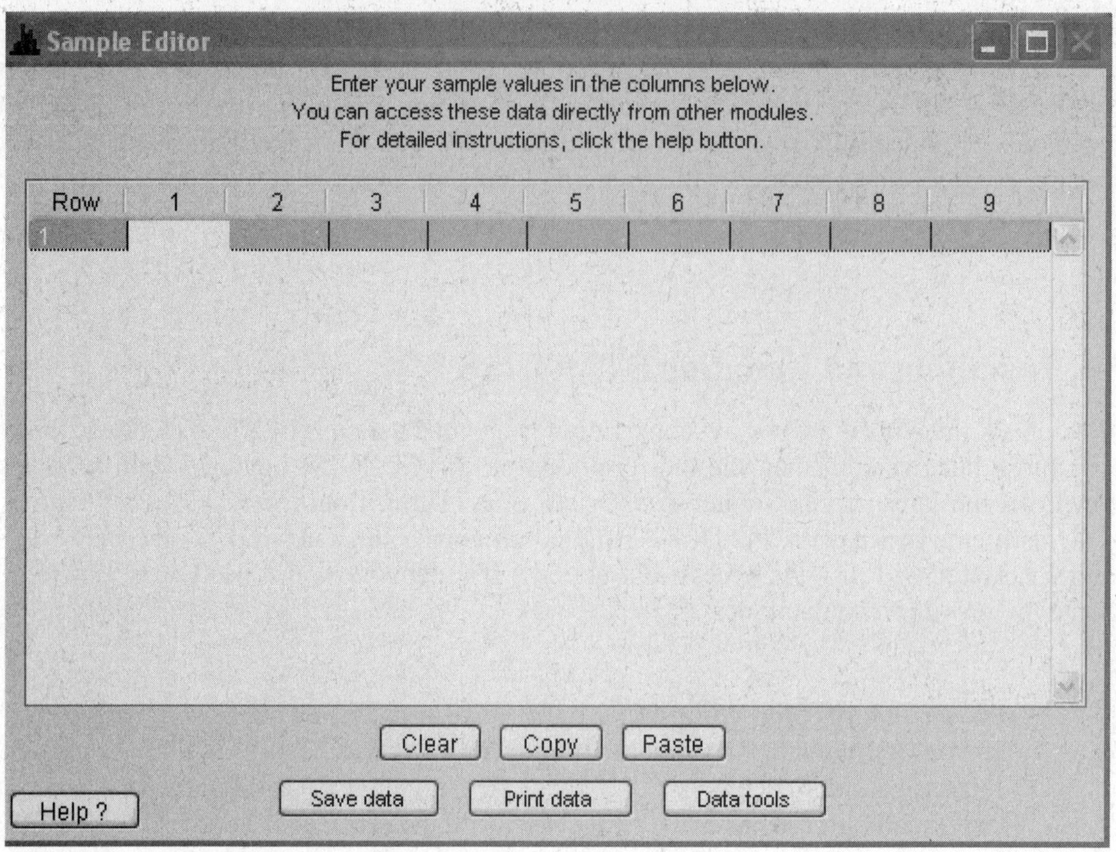

2. Type your first data value in row 1 of column 1, then press the **Enter** key. Then type the second data value and press the **Enter** key again. Proceed to type a sample value followed by pressing the **Enter** key until all of the sample values have been entered. *Note*: If you see that you have made a mistake, simply double-click on the wrong value and make the correction.

There are a few other important features that are available by clicking on the **Data tools** bar. When in the Statdisk data window, if you click on the Data tools bar, you will get the following two menu items:

Edit column titles: Allows you to enter or modify the *names* of the columns of data

Sort data: Allows you to *sort* a column of data (by arranging the data in creasing order).

Delete columns: Allows you delete specific columns.

After entering a large data set, you often want to save it for future use, and you often want to use the data in other modules, such as the modules designed to generate histograms, boxplots, or descriptive statistics (all described in the textbook). Be sure to see Section 1-3 (for saving data) and Section 1-5 (for using *copy* and *paste*).

1-3 Saving Data

After entering a set of data as described above, you can save it for the future by using the following procedure.

STATDISK Procedure for Saving Data

1. After entering all of the values in the Statdisk data window, use the mouse to click on **File** located on the main menu at the top.

   File Edit Analysis Data Datasets Window Help

2. A menu will pop up. Click on the item of **Save As…**.

3. You will now see a dialog box with the title of "Save your STATDISK data set." See the **Save in** box for the location to be used for storing the data set. *Important:* If you want to save your data set in a location different from the default location already shown, you can change the drive and folder as you desire. For example, if you want to save a data set named Heights in a file named Temp, do the following:

   i. Enter "Heights" in the box labeled "File name."

   ii. Click on the symbol ∨ immediately to the right of the window showing the "Save in" file that is current. Select the location to be used for saving the data set.

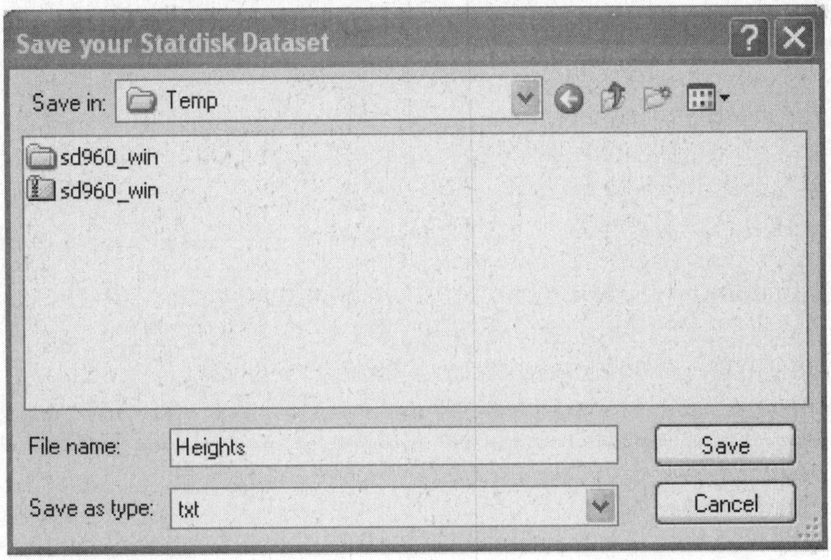

Save your Statdisk Dataset

Save in: Temp

sd960_win

sd960_win

File name: Heights

Save as type: txt

Save

Cancel

1-4 Retrieving Data

STATDISK comes with the Appendix B data sets already stored. Some Exercises in the textbook require that you use some of these data sets. Also, you may want to retrieve a data set that you have previously entered and saved. To retrieve one of the stored data sets, follow these easy steps.

1. Click on the main menu item of **Datasets**.

2. Select one of the Appendix B data sets. Click on the desired name. For example, to select the Homeruns data set, use the mouse to scroll down to the name of **Homeruns**, then click on that name. The data set will be inserted in the Statdisk data window, as shown below. (The complete column titles of McGwire, Sosa, and Bonds can be seen by clicking on **Data tools** and selecting the option of **Edit column titles**.)

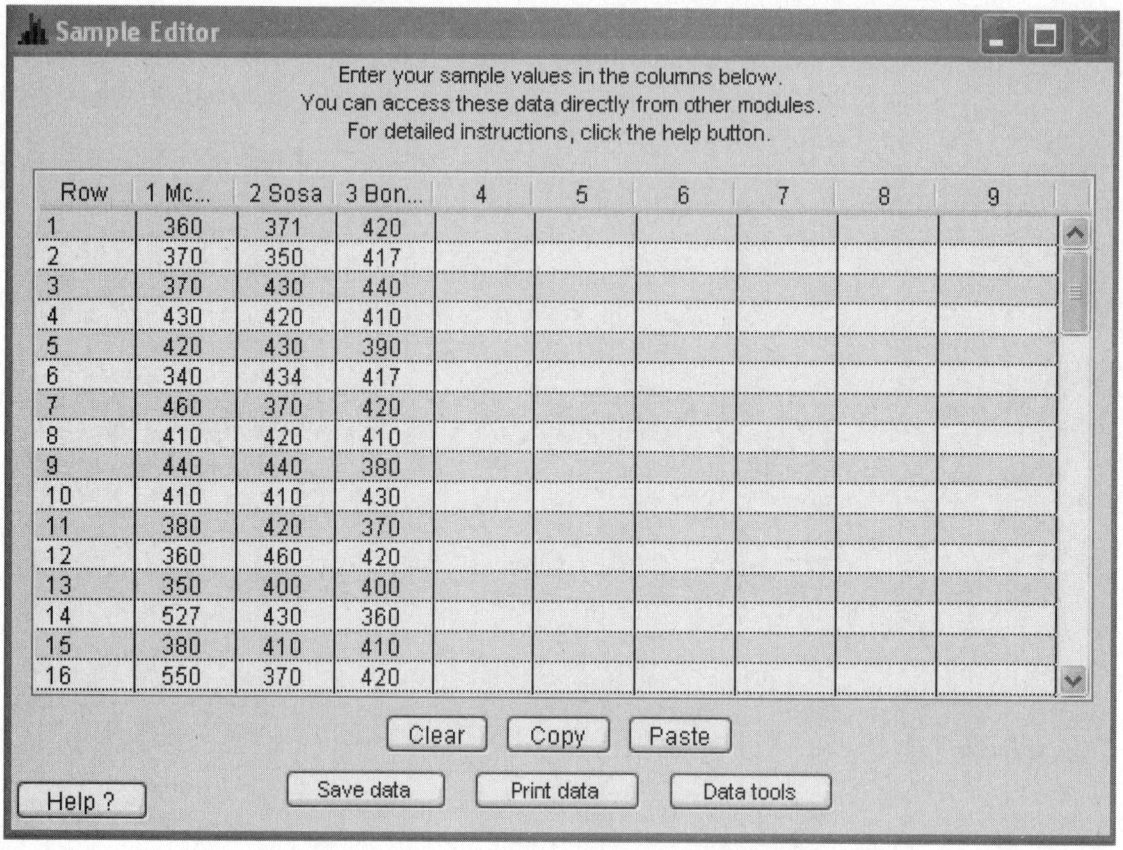

Sample Editor

Enter your sample values in the columns below.
You can access these data directly from other modules.
For detailed instructions, click the help button.

Row	1 Mc...	2 Sosa	3 Bon...	4	5	6	7	8	9
1	360	371	420						
2	370	350	417						
3	370	430	440						
4	430	420	410						
5	420	430	390						
6	340	434	417						
7	460	370	420						
8	410	420	410						
9	440	440	380						
10	410	410	430						
11	380	420	370						
12	360	460	420						
13	350	400	400						
14	527	430	360						
15	380	410	410						
16	550	370	420						

Clear Copy Paste

Help ? Save data Print data Data tools

Although only 16 values are visible in each of the three columns, the other data values can be viewed by scrolling down the list. *Note:* STATDISK allows up to nine columns, and some of the Appendix B data sets require more than nine columns, so they are stored in two parts. In such cases, individual columns can be copied and moved so that desired columns can be combined as desired.

1-5 Copy/Paste

The Copy and Paste feature is an extremely important tool of many different software applications, including word processors and spreadsheets. You should clearly understand the following.

- **After entering or retrieving a data set and using the *Copy* command, the data set will remain available for use until you use *Copy* for a new data set, or until you exit the program.**

- **After using the *Copy* command, go to the module where you want to use the data set, then click on the *Paste* button in the window for that application.**

See Section 1-9 for procedures allowing you to copy data sets between STATDISK and other applications, such as Minitab or Excel or Word. The procedure below illustrates the usefulness of the Copy/Paste feature within STATDISK.

STATDISK Procedure for Using Copy and Paste

1. Enter the data in the Statdisk data window (as described in Section 1-2 above) or retrieve a stored data set into the Statdisk data window).

2. Click on the **Copy** button located at the bottom of the Statdisk data window.

3. You will see a pop-up window that gives you the choice of copying *all* of the columns or individual columns. For example, the screen below shows that columns 1 and 3 have been designated for copying. Click on the **Copy** button at the bottom.

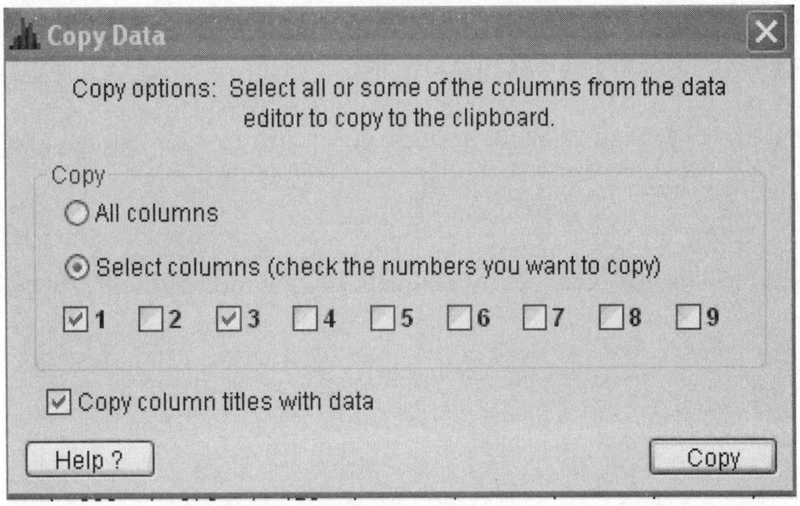

Copy Data

Copy options: Select all or some of the columns from the data editor to copy to the clipboard.

Copy

All columns

Select columns (check the numbers you want to copy)

1 2 3 4 5 6 7 8 9

Copy column titles with data

Help ? Copy

Here are some options for using the copied columns of data:

- *Isolate desired columns of data:* Clear the Statdisk data window, then click the **Paste** button to insert only those columns that were previously copied.

- *Combine desired columns of data:* Open another data set in the Statdisk data window, then click the **Paste** button to insert the copied columns. For example, suppose that you want to combine columns of data from different data sets, such as the weights of bears and the weights of men. Weights of bears are listed in the Bears data set, and weights of men are listed in the MHealth data set. (Both data sets are found in Appendix B of the Triola textbook.) The Copy/Paste feature can be used to combine those data sets. Simply copy the bear weights into the MHealth data set, so that the weights of the bears and the weights of men are included in the same Statdisk data window.

1-6 Editing and Transforming Data

It is easy to edit a data set while in the Statdisk data window.

- *Delete* an entry by clicking on it and using the **Del** key to remove it.

- *Insert* an entry by typing it in an empty cell at the bottom of the list.

Data may also be *transformed* with operations such as adding a constant, multiplying by a constant, or using the functions of adding, subtracting, multiplying, dividing, raising to a power, or stripping away the decimal part of data values. For example, if you have a data set consisting of temperatures on the Fahrenheit scale (such as the Body Temperature data set in Appendix B of the textbook) and you want to transform the values to the Celsius scale, you can use the equation

$$C = \frac{5}{9}(F - 32)$$

STATDISK Procedure for Transforming Data

1. First enter the data in one or more columns of the Statdisk data window.

2. Click on the main menu item of **Data**.

3. Click on **Sample Transformations** to get a window like the one shown below.

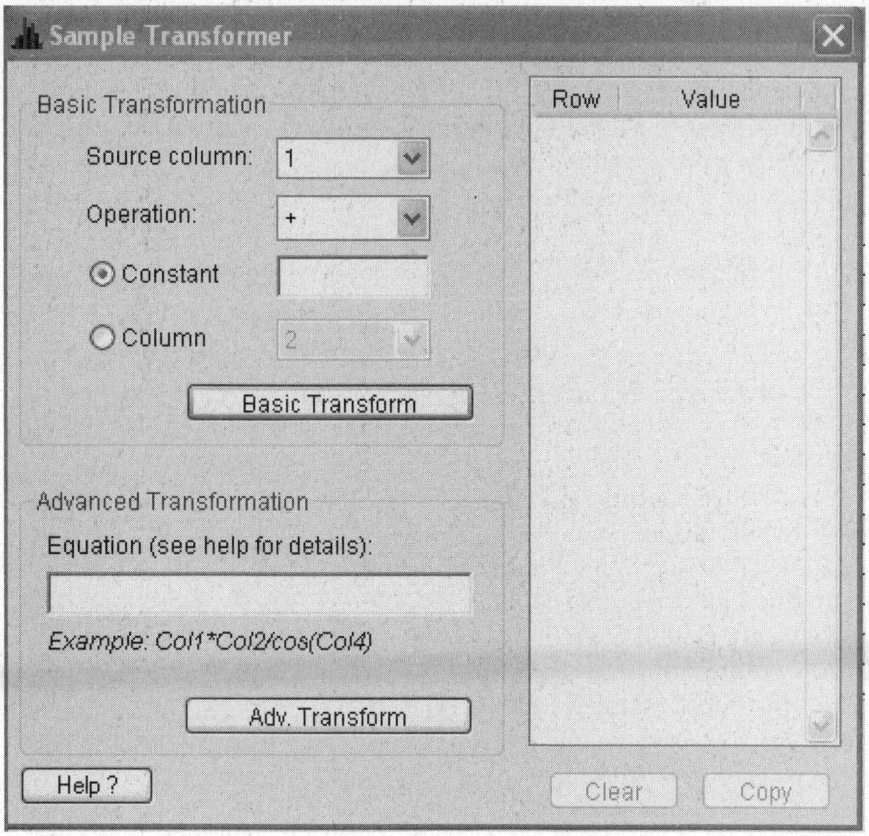

Sample Transformer
Basic Transformation
Source column: 1
Operation: +
Constant
Column 2
Basic Transform
Advanced Transformation
Equation (see help for details):
*Example: Col1*Col2/cos(Col4)*
Adv. Transform
Help ?
Row Value
Clear Copy

4. For the Source column, select the column of data to be transformed.

5. For the operation, select the desired operation from the list of available options.

6. Select "Constant" to perform the operation with the same value, or select "Column" to perform the operation with corresponding values in two different columns.

7. Click on **Basic Transform** and the transformed values will appear in the column at the extreme right. The result can be copied to the STATDISK data window.

The "Advanced Transformation" feature can be used for more advanced transformations, such as those involving absolute values, logarithms, or the sine function. For more details of the procedure for transforming data, select **Data** from the main menu, then select **Sample Transformations**, then click on the **Help** button.

1-7 Printing Screens and Data Sets

After you have obtained results from STATDISK, such as a graph or a listing of statistics, you can print those results. Simply click on the **Print** button. The Print button might not be visible or enabled until the results have been obtained by clicking on the **Evaluate** button first.

Including a Name With the Printout A problem existed when several students were all printing at the same time in a computer lab. Students were not able to identify which printout was theirs. STATDISK now allows you to enter a label or name so that printed results can be identified. When clicking on the **Print** button for the first time in a session, you will see a pop-up window. If you want to have your name included with printed results, click on the **Open Preferences** button, then be sure that the **On** button is selected in the Printing section at the bottom. Be sure to enter your name in the box with the message of "This label will be printed." The time and date can also be included.

It was not possible to test all printing configurations and all printers. If you find that pressing the Print button does not produce a satisfactory result, press the Alt-PrintScreen key to copy the screen to a clipboard so that the screen can then be pasted to any document editor (such as MS Word) or MS Paint. Another option is to use a screen capture program, such as one available from www.any-capture.com.

If you click on the Print Data button in the Statdisk data window, STATDISK will print all of the data, not just those visible on the screen. If you have a large STATDISK data set that you would like to print, it is recommended that you move it to a word processor, such as MS Word, where you can reconfigure the data for more efficient printing. Use the following procedure.

Printing a Large Data Set in a Word Processor

1. With the data set listed in the Statdisk data window, click on the **Copy** button.

2. In your word processor, click on **Edit**, then **Paste**. The entire list of values will be in your word processing document where you can configure them as you please. Instead of printing a data set of 1000 values with one column covering many pages, use many values in each row so that fewer pages will be printed. Environmentalists will be very thankful.

1-8 Closing Windows and Exiting

Closing Windows: To minimize clutter, *close* windows after they are no longer needed. Close windows by clicking on the small box labeled × that is located in the upper right corner of the window. (Clicking on the box with the symbol – will cause the window to be hidden, but it continues to remain open and available for recall.)

Exiting STATDISK: Had enough for now? To exit or quit the STATDISK program, click on the × located in the extreme upper right corner. Another way to exit STATDISK is to click on **File**, then click on **Quit**.

1-9 Exchanging Data with Other Applications

There may be times when you want to move data from STATDISK to another application (such as Excel or Minitab or Word) or to move data from another application to STATDISK. Instead of manually retyping all of the data values, you can usually transfer the data set directly. Given below are two ways to accomplish this.

Method 1: Use Copy and Paste

1. With the data listed in the Statdisk data window, click on the **Copy** button.

2. A pop-up window will give you the choice of copying *all* of the columns or individual columns. For example, the screen below shows that columns 1 and 3 have been designated for copying. Click on the **Copy** button at the bottom.

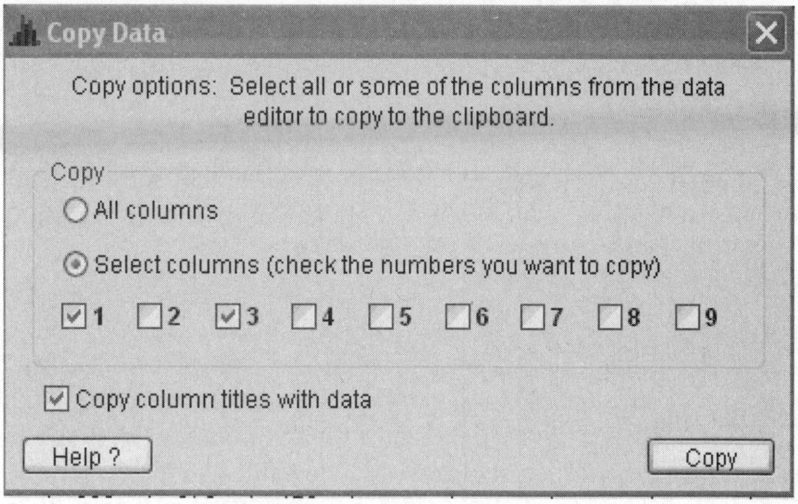

Copy Data

Copy options: Select all or some of the columns from the data editor to copy to the clipboard.

Copy

All columns

Select columns (check the numbers you want to copy)

1 2 3 4 5 6 7 8 9

Copy column titles with data

Help ? Copy

3. You can now paste the data to other applications. For example, load Minitab, click on the data cell in the first row and first column, then click on Minitab's **Edit** and **Paste** items. The copied column(s) will reappear in Minitab, and you can proceed to use that software package with the STATDISK-generated data.

Method 2: Use Text Files

1. In the software program containing the original set of data, create a text file of the data. (STATDISK files are already text files.)

2. STATDISK and most other major applications allow you to import the text file that was created. (To import a text file into STATDISK, select **File**, then **Open**. Enter the location of the text file in the "Look in" box, then click the **Open** button.)

CHAPTER 1 EXPERIMENTS: STATDISK Fundamentals

1-1. ***Entering Sample Data*** When first experimenting with procedures for using STATDISK, it's a good strategy to use a small data set instead of one that is large. If a small data set is lost, you can easily enter it a second time. In this experiment, we will enter a small data set, save it, retrieve it, and print it.

98.6 98.6 98.0 98.0 99.0

a. Open STATDISK and enter the above sample temperatures. (See the procedure described in Section 1-2 of this manual/workbook.)

b. Save the data set using the file name of SMALL. See the procedure described in Section 1-3 of this manual/ workbook.

c. Print the data set.

d. Exit STATDISK, then reload it and retrieve the file named SMALL. Save another copy of the same data set using the file name of SMALL2. Print SMALL2 and include the title at the top. (That is, use a title of SMALL2 instead of the default.)

1-2. ***Retrieving Data*** The data set Bears is already stored in STATDISK. It contains nine columns with 54 values in each column. The data are listed in Appendix B of the Triola textbook. Open that file and print the data.

1-3. ***Using Copy/Paste*** Enter the sample values 1, 2, 3, 10, 20 and use **Data/Sample Transformations** to add 5 to each value, then use Copy/Paste to copy the results to the second column of the Statdisk data window. Then click **Data**, select **Descriptive Statistics**, select column 2, and click **Evaluate** and obtain a printed copy of the resulting screen display. The resulting statistics will be described in the textbook.

1-4. ***Using Copy/Paste*** The data sets Mhealth (for male health measurements) and Fhealth (for female health measurements) are stored in STATDISK. Use Copy/Paste to create a Statdisk data window that includes only these two columns: (1) heights of the males and (2) heights of the females. Obtain a printed copy of that STATDISK data window

1-5. ***Editing Data*** The data set Homeruns includes the 73 distances of the homeruns hit by Barry Bonds when he broke a major baseball record.

a. Open the data set Homeruns, then find the value of the mean of the Bonds distances by using **Data** and **Descriptive Statistics.** Enter the value of the mean.________

b. Go back to Statdisk data window and change the fifth value from 390 to 3900. Repeat part a and record the new value of the mean.__________

c. Did the mean change much when the fifth value was changed from 390 to 3900?

____

1-6. ***Generating Random Data*** In addition to entering or retrieving data, STATDISK can also *generate* data sets. In this experiment, we will use STATDISK to simulate the rolling of a pair of dice 500 times. Select **Data** from the main menu bar, then select **Dice Generator**. For the sample size, enter 500 (for 500 rolls), enter 2 for the number of dice, enter 6 for the number of sides, then click on **Generate**. Examine the displayed totals and count the number of times that 7 occurs. Record the result here: ___________

1-7. ***Transforming Data*** Experiment 1-1 results in saving a sample of body temperatures in degrees Fahrenheit. Retrieve that data set, then proceed to transform the temperatures to the Celsius scale. (See Section 1-6 in this manual/workbook.) After obtaining the Celsius temperatures, use **Data** and **Descriptive Statistics** to find the value of the mean, and record the result here:_______________

1-8. ***Retrieving and Transforming Data*** Open the STATDISK data set Bears, which includes the weights (in pounds) of a sample of bears. To convert the weights to kilograms, multiply them by 0.4536. Use STATDISK to convert the weights from pounds to kilograms. In the space below, write the weights (in kilograms) of the first five bears.

____

1-9. ***Entering and Saving a Data Set*** Listed below are the ages of actresses and actors at the time they won Oscars. Enter the ages of the winning actresses in the first column and name that column Actresses. Enter the ages of the winning actors in the second column and name that column Actors. Save the data set with the name OSCAR. Print the data set.

Actresses

22	37	28	63	32	26	31	27	27	28	30	26
29	24	38	25	29	41	30	35	35	33	29	38
54	24	25	46	41	28	40	39	29	27	31	38
29	25	35	60	43	35	34	34	27	37	42	41
36	32	41	33	31	74	33	50	38	61	21	41
26	80	42	29	33	35	45	49	39	34	26	25
33	35	35	28								

Actors

44	41	62	52	41	34	34	52	41	37	38	34
32	40	43	56	41	39	49	57	41	38	42	52
51	35	30	39	41	44	49	35	47	31	47	37
57	42	45	42	44	62	43	42	48	49	56	38
60	30	40	42	36	76	39	53	45	36	62	43
51	32	42	54	52	37	38	32	45	60	46	40
36	47	29	43								

2 Graphing Data

Important note: The topics of this chapter require that you use STATDISK to enter data, retrieve data, save files, and print results. These functions are covered in Chapter 1 of this manual/ workbook. Be sure to understand these functions before beginning this chapter.

In this chapter, we learn how to use STATDISK as a tool for graphing and summarizing data. Let's consider the data from Experiment 1-9 in Chapter 1 of this manual/ workbook. The data are reproduced below. If Experiment 1-9 was successfully completed, the data set is saved with the name of Oscar and it can be retrieved.

Actresses

22	37	28	63	32	26	31	27	27	28	30	26
29	24	38	25	29	41	30	35	35	33	29	38
54	24	25	46	41	28	40	39	29	27	31	38
29	25	35	60	43	35	34	34	27	37	42	41
36	32	41	33	31	74	33	50	38	61	21	41
26	80	42	29	33	35	45	49	39	34	26	25
33	35	35	28								

Actors

44	41	62	52	41	34	34	52	41	37	38	34
32	40	43	56	41	39	49	57	41	38	42	52
51	35	30	39	41	44	49	35	47	31	47	37
57	42	45	42	44	62	43	42	48	49	56	38
60	30	40	42	36	76	39	53	45	36	62	43
51	32	42	54	52	37	38	32	45	60	46	40
36	47	29	43								

In Chapter 2 of the textbook, we begin with the construction of frequency distributions, followed by the construction of histograms. We reverse the order of presentation in this manual/workbook. We begin with histograms.

2-1 Histograms

The textbook describes the manual construction of a histogram, but STATDISK can be used to automatically generate this important graph. The basic approach is to get the data listed in the Statdisk data window, then use the Histogram module, where the histogram graph is generated. When using STATDISK's Histogram program, you have the option of simply accepting default settings, or you can set your own limits on the classes. If you choose to set your own limits, you must understand the definition of *class width*. In the textbook, we define class width as follows:

> **Class width** is the difference between two consecutive lower class limits or two consecutive lower class boundaries.

As an example, see the following frequency distribution obtained from the ages of actresses listed above. *The class width is 10* (the difference between the consecutive lower class limits of 21 and 31).

Age of Actress	Frequency
21-30	28
31-40	30
41-50	12
51-60	2
61-70	2
71-80	2

Procedure for Generating a Histogram

1. Enter or retrieve a set of sample data using one of these procedures:

   - **Manual entry of data:** Values can be entered in the Statdisk data window.

   - **Retrieve a data set from those included in Appendix B:** Click on the main menu item of **Datasets** and proceed to select one of the listed data sets.

   - **Retrieve a data set that you created:** Use **File/Open** as described in Section 1-4 of this manual/workbook.

2. Click on **Data** in the main menu bar at the top.

3. Click on **Histogram**.

4. In the Histogram window, first select the column to be used. The default is column 1, and it can be changed to any column number between 1 and 9.

5. Click **Plot** to allow STATDISK to automatically generate a histogram using default settings. (You have the option of changing the class width and starting point of the first class. You also have the option of having a vertical scale that uses relative frequencies instead of actual frequency counts. Click OK after making the desired changes.)

As an example, see the STATDISK display on the following page. The histogram is constructed using the preceding list of the ages of actresses. Instead of using the default settings, we clicked on the **User defined** button and entered 10 for the class width and 21 as the starting point. These entries correspond to the frequency table shown above. (See the preceding frequency distribution table and verify that the class width is 10 and the lower limit of the first class is 21.)

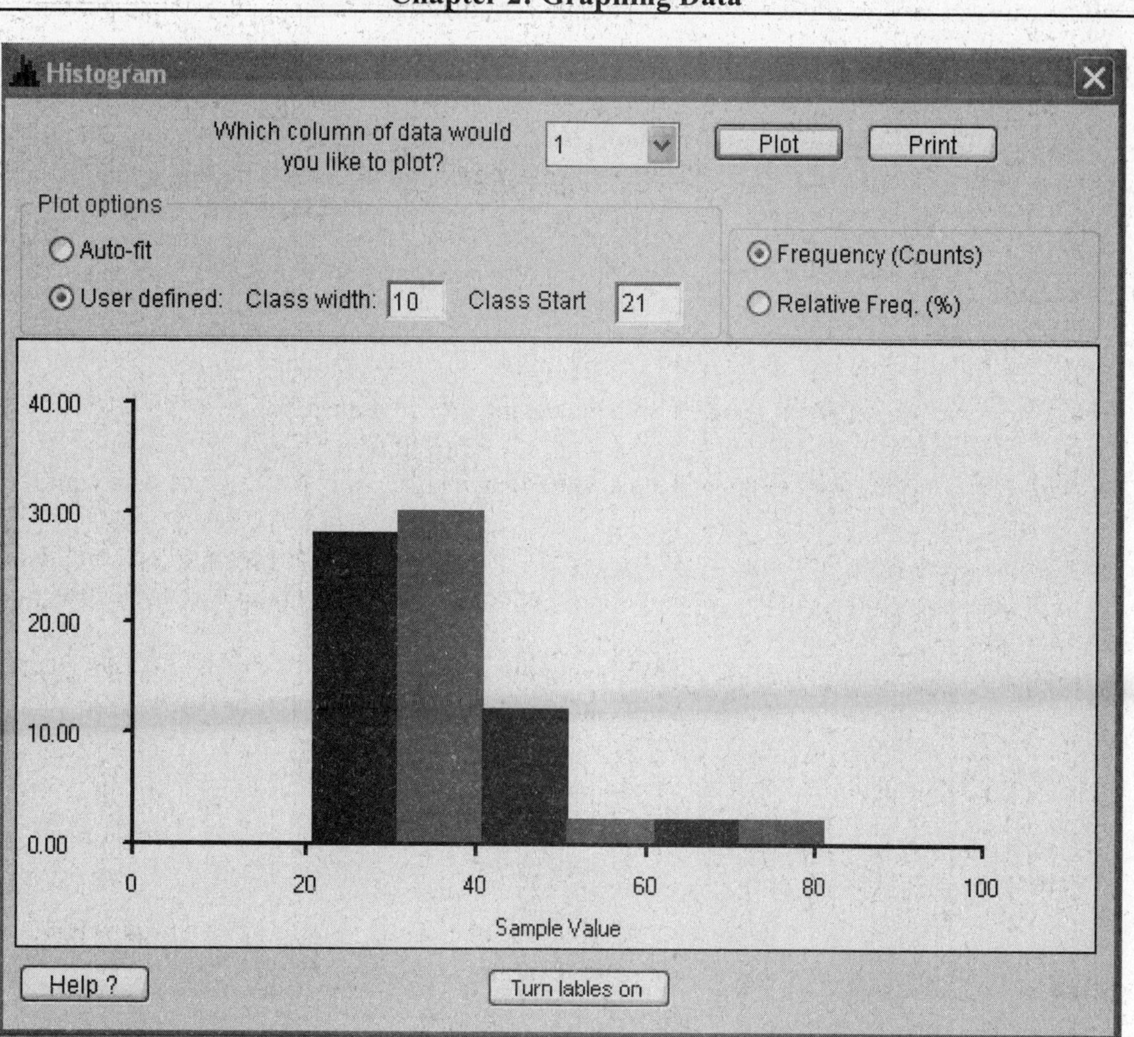

The histogram gives us insight into the nature of the *distribution*. In later chapters, it often becomes important to determine whether sample data appear to come from a population with a normal distribution. A "normal distribution" will be discussed later, but for now we can consider it to be a distribution with a histogram that is roughly bell–shaped. For now, simply examine the histogram and make a judgment about whether it appears to be approximately bell-shaped. If we examine the STATDISK histogram shown above, we can see that the distribution does not appear to be very bell–shaped, so that a requirement of a normal distribution does not appear to be satisfied for this data set. (STATDISK can also generate *normal quantile plots,* which are helpful for determining whether distributions are normal, and those graphs are discussed in Chapter 6 of this manual/workbook.)

2-2 Frequency Distributions

In designing STATDISK, we did not include a specific menu item for generating a frequency distribution from a list of raw data, but frequency distributions can be obtained by using the ability of STATDISK to generate histograms. If you want to use STATDISK to construct a frequency distribution, enter your own starting point and class width (based on the range of values and the minimum value).

See the above STATDISK display and note that there is a "**Turn labels on"** button. If you click on that button, you get the labels consisting of frequency counts above the bars of the histogram as shown below. Knowing that the class width is 10 and the lower limit of the first class is 21, we can see that the first class is 21-30 and it has a frequency of 28. Similarly, the next class is 31-40 and it has a frequency of 30 as shown in the display below.

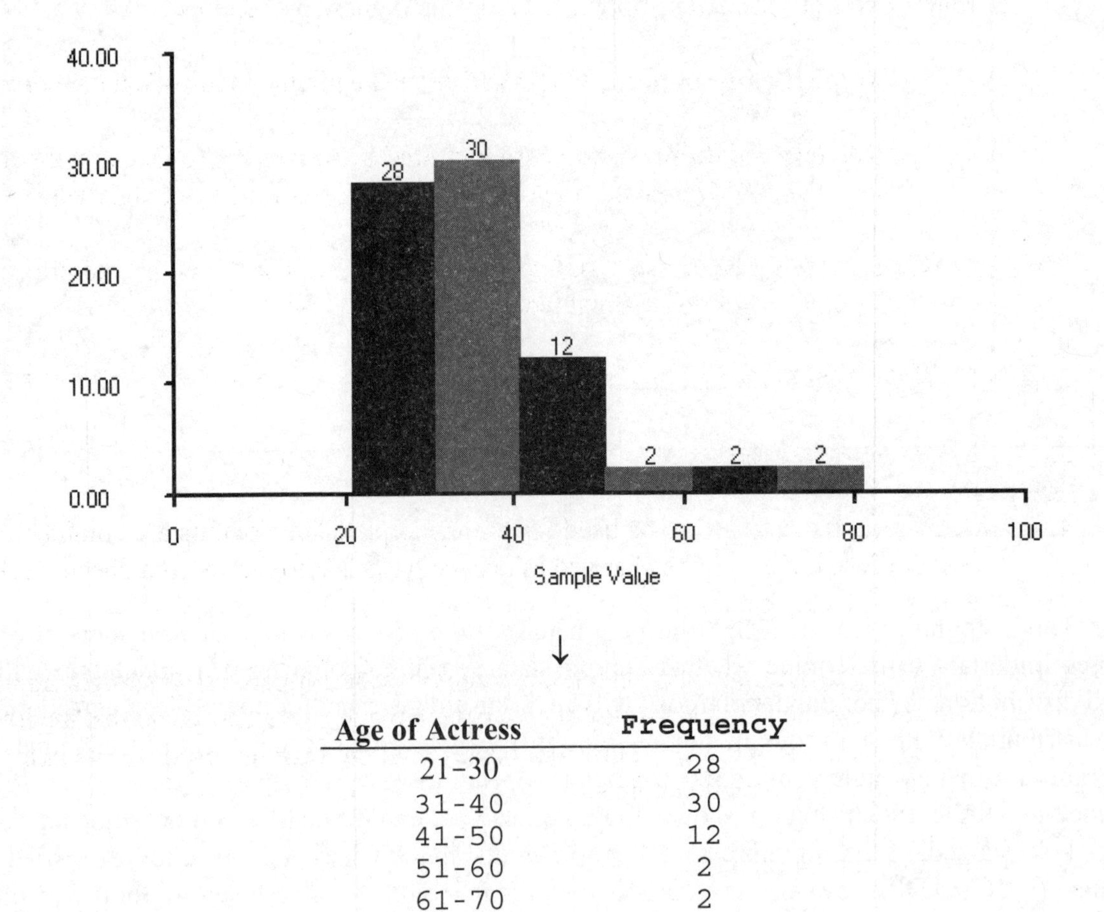

↓

Age of Actress	Frequency
21-30	28
31-40	30
41-50	12
51-60	2
61-70	2
71-80	2

Technical note: STATDISK is designed so that a sample value falls into a particular class if it is equal to or greater than the lower class limit and less than the upper class limit.

2-3 Scatterplots

The textbook describes a scatterplot (or scatter diagram) as a plot of paired (x, y) data with a horizontal x-axis and a vertical y-axis. The data are paired in a way that matches each value from one data set with a corresponding value from a second data set. A scatterplot can be very helpful in seeing a relationship between two variables.

Procedure for Generating a Scatterplot

To use STATDISK for generating a scatterplot, you must have a collection of *paired* data listed in the Statdisk data window.

1. Enter or retrieve a set of sample data using one of these procedures:

   - **Manual entry of data:** Values can be entered in the Statdisk data window.

   - **Retrieve a data set from those included in Appendix B:** Click on the main menu item of **Datasets** and proceed to select one of the listed data sets.

   - **Retrieve a data set that you created:** Use **File/Open** as described in Section 1-4 of this manual/workbook.

2. Click on **Data** in the main menu bar at the top.

3. Click on **Scatterplot**.

4. Select the two columns to be used for the scatterplot. The defaults are column 1 and column 2, and they can be changed as desired. (Click a box to insert a check mark or to remove a check mark.)

5. If the "Regression line" box is left checked, the graph will include a straight line that best fits the points. In the display shown below, the box was clicked so that the line is not included. Also, the "Turn on labels" button can be used to include the coordinates of the plotted points.

6. Click on the **Plot** button.

If we use the paired ages of actresses and actors listed earlier in this chapter, we get the scatter diagram shown below. Based on the pattern of the points, we can conclude that there does not appear to be a relationship between the ages of Oscar-winning actresses and the ages of Oscar-winning actors, given that the ages are paired by the year in which the Oscars were won. Such relationships (or *correlations*) will be discussed at much greater length in the Correlation and Regression chapter.

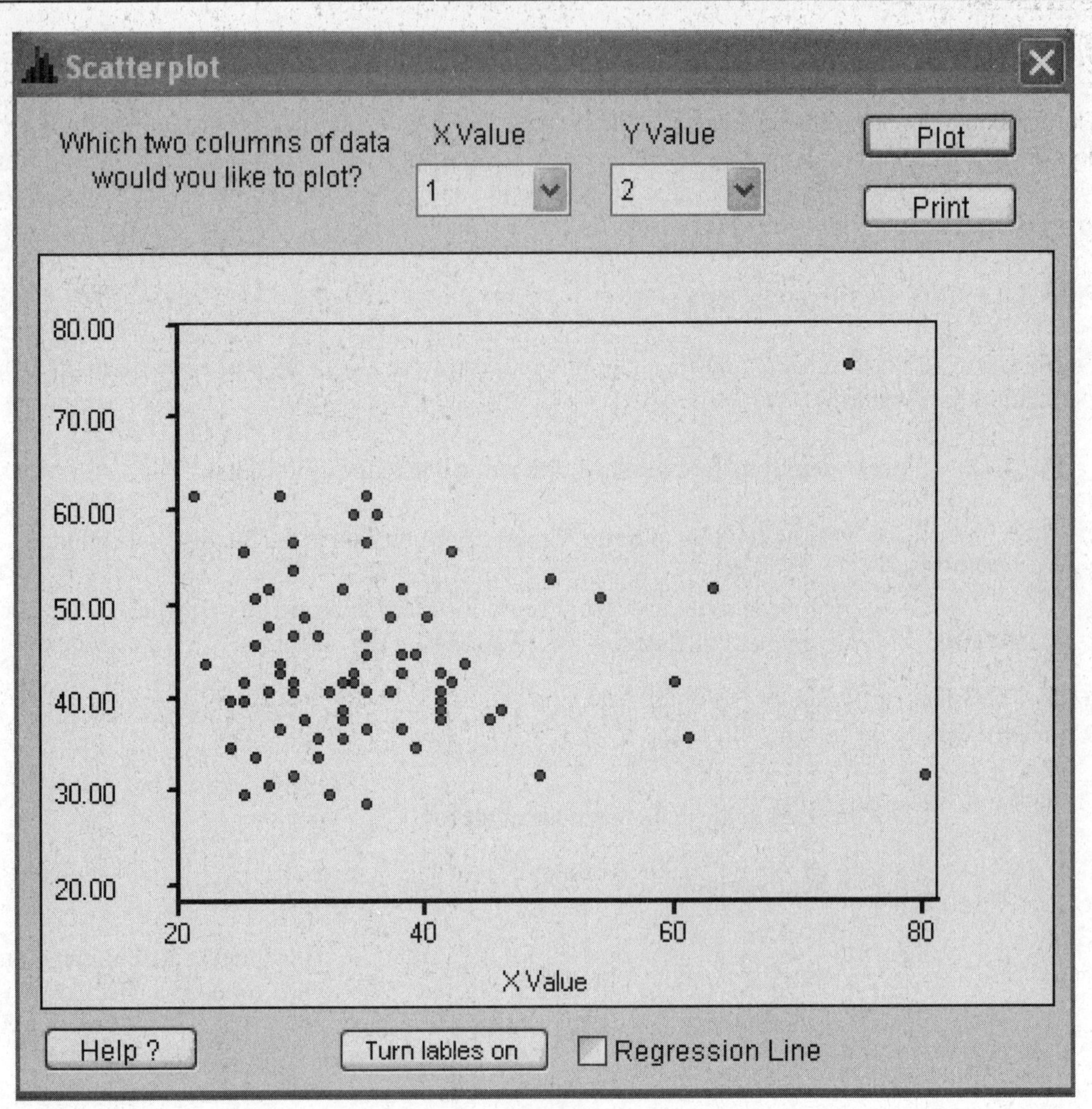
Scatterplot
Which two columns of data would you like to plot?
X Value
1
Y Value
2
Plot
Print
80.00
70.00
60.00
50.00
40.00
30.00
20.00
20
40
60
80
X Value
Help ?
Turn lables on
Regression Line

2-4 Pie Charts

The textbook makes the point that pie charts are generally poor as effective graphs for depicting data. STATDISK does include an option for generating pie charts.

Procedure for Generating a Pie Chart

1. Begin by entering the category names as text in column 1 of the STATDISK data window.

2. Next, enter the corresponding frequency counts in column 2.

3. Select **Data** from the main menu at the top, then select **Pie Chart**.

4. Click on the **Chart** button and the pie chart will be displayed.

2-5 Sorting Data

To *sort* data is to arrange them in order. There are several cases in which it becomes necessary to rearrange a data set so that the values are in order, ascending from low to high.

Procedure for Sorting Data

1. Enter or retrieve a set of sample data so that the sample values are listed in the Statdisk data window.

2. Click on the **Data tools** bar located near the bottom of the STATDISK data window.

3. Proceed to select the option of **Sort a column**.

4. Select the column to be sorted, then click the **Sort** button.

Using Sort to Identify Outliers The sort feature is useful for identifying outliers. When analyzing data, it is important to identify outliers because they can have a dramatic effect on many results. It is usually difficult to recognize an exceptional value when it is buried in the middle of a long list arranged in a random order, but *outliers become much easier to recognize with sorted data, because they will be found either at the beginning or end.* To identify outliers, simply sort the data, then examine the lowest and highest values to determine whether they are dramatically far from almost all of the other sample values. If we plan to further analyze the sample data, we should be aware of any outliers, because they might dramatically affect some of our results.

CHAPTER 2 EXPERIMENTS: Graphing Data

2–1. ***Histogram*** Use the 40 BMI (body mass index) indices of women from the data set Fhealth (female health) and print a histogram. Instead of using the default values, use a class width of 6.0 and using a lower class limit of 15.0. Does the histogram appear to be approximately bell-shaped?

____

Use STATDISK to obtain a frequency distribution with a class width of 6.0 and a lower class limit of 15.0 for the first class. Enter the frequency distribution here:

2–2. ***Histogram*** Use the 40 BMI (body mass index) indices of men from the data set Mhealth (male health) and print a histogram. Instead of using the default values, use a class width of 6.0 and using a lower class limit of 15.0. Does the histogram appear to be approximately bell-shaped?

____

Use STATDISK to obtain a frequency distribution with a class width of 6.0 and a lower class limit of 15.0 for the first class. Enter the frequency distribution here:

Compare the histogram from this experiment to the one obtained in Experiment 2-1. Do females and males have BMI values that appear to be dramatically different? Explain.

____

____

____

Can a scatterplot be constructed by using the 40 BMI values of females and the 40 BMI values of males? Does it make sense to construct a scatterplot in this case? Explain.

____

____

2–3. ***Scatterplots*** Use the data set Mhealth (male health) to print the scatterplot for the paired waist and weight measurements for the sample of 40 men. Does there appear to be a relationship between waist sizes and weights of males? Explain.

____

____

____

2–4. ***Scatterplots*** Use the data set Fhealth (female health) to print the scatterplot of the paired systolic and diastolic blood pressure measurements for the sample of 40 women. Does there appear to be a relationship between systolic blood pressure and diastolic blood pressure? Explain.

____

____

____

2–5. ***Scatterplots*** Use the data set Bears to print the scatterplot of the paired length and weight measurements. Does there appear to be a relationship between lengths of bears and their weights? Explain.

____

____

____

2–6. ***Scatterplots*** Use the data set Bears to print the scatterplot of the paired chest size and weight measurements. Does there appear to be a relationship between chest sizes of bears and their weights? Explain.

____

____

____

Compare the scatterplot obtained in this experiment to the one obtained in Experiment 2-5. Which association is stronger: the length/weight association or the chest/weight association? Explain your choice.

____

____

____

2–7. ***Frequency Distribution*** Shown below is a STATDISK–generated histogram representing the ages (n years) of eastbound stowaways on the Queen Mary. Use the displayed histogram to construct a table representing the frequency distribution, and identify the value of the class width.

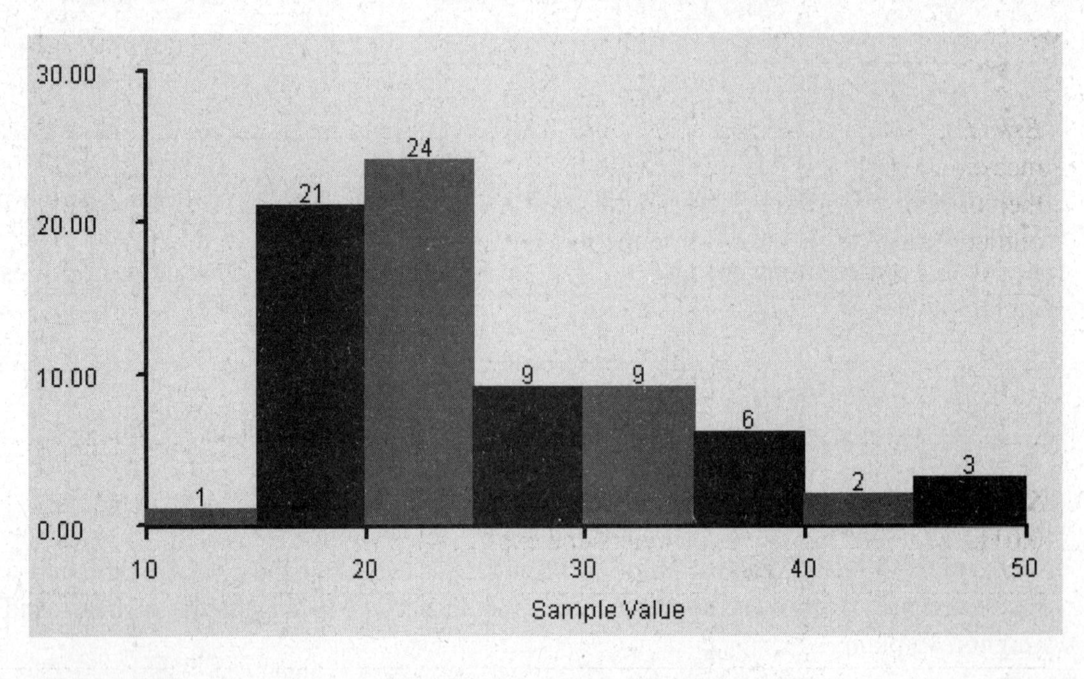

Class width: __________ Enter the frequency distribution here:

2–8. ***Frequency Distribution*** The accompanying STATDISK–generated histogram represents the times (in seconds) of runners who finished the New York Marathon. Use the displayed histogram to construct a table representing the frequency distribution, and identify the value of the class width. Enter them on the top of the following page.

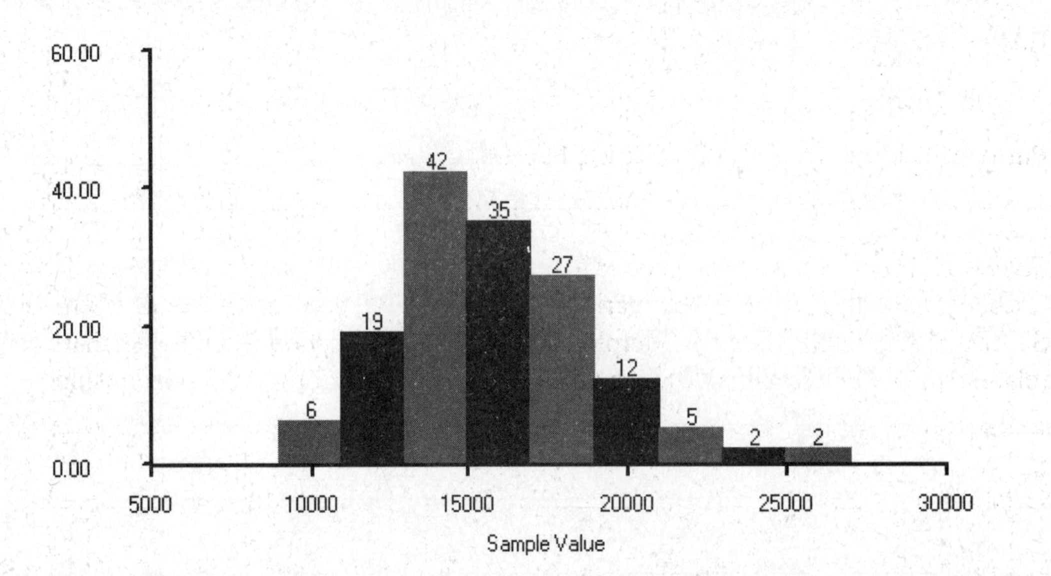

Class width: __________ Enter the frequency distribution here:

2–9. ***Effect of Outlier*** In Experiment 2–1 we obtained a printout of a histogram for the BMI measurements of 40 females. Change the first BMI value from 19.6 to the outlier of 196, then obtain the histogram and print it. How is the histogram affected by the presence of the outlier? Does the outlier disguise the true nature of the distribution of the data?

2–10. ***Sorting Data*** Open the STATDISK file Cans and *sort* the axial loads of the cans that are 0.0111 in. thick. Identify any outliers and explain why they were selected.

2-11. ***Exploring Distributions*** Open each of the indicated data sets and obtain a printed copy of a STATDISK–generated histogram. Examine the histogram and described its general shape. Determine whether the shape of the distribution is approximately bell-shaped.

Cotinine levels of smokers (STATDISK file Cotinine):

Body temperatures of healthy adults from 12 AM on day 2 (STATDISK file Bodytemp):

Pulse rates of females (STATDISK file Fhealth):

2–12. ***Scatterplots*** Now it's time to be creative. Using any data set from Appendix B in the textbook, identify two paired variables that you suspect are related, then obtain a STATDISK printout of the scatterplot. Does the graph support your belief that there is a relationship? What feature of the graph suggests that there is or is not a relationship?

Statistics for Describing, Exploring, and Comparing Data

Important note: The topics of this chapter require that you use STATDISK to enter data, retrieve data, save files, and print results. These functions are covered in Chapter 1 of this manual/workbook. Be sure to understand these functions before beginning this chapter.

Important Characteristics of Data

When describing, exploring, and comparing data sets, the following characteristics are usually extremely important:

1. **Center:** Measure of center, which is a representative or average value that gives us an indication of where the middle of the data set is located

2. **Variation:** A measure of the amount that the values vary among themselves

3. **Distribution:** The nature or shape of the distribution of the data, such as bell-shaped, uniform, or skewed

4. **Outliers:** Sample values that are very far away from the vast majority of the other sample values

5. **Time:** Changing characteristics of the data over time

In Chapter 2 of this manual/workbook we addressed the characteristics of distribution and outliers by using histograms and sorted lists of data. In this chapter we address the important measures of center and variation.

In Chapter 2 of this manual/workbook, we used the ages of Oscar winners at the time they received their awards. We will use this same data set in this chapter. If Experiment 1-9 was successfully completed, this data set is saved with the name of OSCAR.

Actresses

22	37	28	63	32	26	31	27	27	28	30	26
29	24	38	25	29	41	30	35	35	33	29	38
54	24	25	46	41	28	40	39	29	27	31	38
29	25	35	60	43	35	34	34	27	37	42	41
36	32	41	33	31	74	33	50	38	61	21	41
26	80	42	29	33	35	45	49	39	34	26	25
33	35	35	28								

Actors

44	41	62	52	41	34	34	52	41	37	38	34
32	40	43	56	41	39	49	57	41	38	42	52
51	35	30	39	41	44	49	35	47	31	47	37
57	42	45	42	44	62	43	42	48	49	56	38
60	30	40	42	36	76	39	53	45	36	62	43
51	32	42	54	52	37	38	32	45	60	46	40
36	47	29	43								

3-1 Measures of Center and Variation

Important measures of center and variation can be obtained by using STATDISK's "Explore Data" or "Descriptive Statistics" features. The Explore Data feature is recommended because it provides much more information. To explore a list of data, follow these steps.

1. Enter or retrieve a set of sample data using one of these procedures:

    - **Manual entry of data:** Values can be entered in the Statdisk data window.
    - **Retrieve a data set from those included in Appendix B:** Click on the main menu item of **Datasets** and proceed to select one of the listed data sets.
    - **Retrieve a data set that you created:** Use **File/Open** as described in Section 1-4 of this manual/workbook.

2. Click on **Data** in the main menu bar at the top.

3. Click on **Explore Data.**

4. Select the column to be used for the calculations and graphs. The default of column 1 can be changed to any column number between 1 and 9.

5. Click on the **Evaluate** button.

As an example, consider the ages of actresses at the time they won Oscars for the Best Actress category. Those ages are listed on the preceding page. Assume that the ages of actresses are listed in column 1 of the STATDISK data window, and the above procedure is followed. The result will be as shown on the following page.

From the display on the next page we see that there are $n = 76$ sample values, the sample mean is 35.7 years (rounded), the median is 33.5 years, the midrange is 50.5 years. The value of "RMS" is the value of the *root mean square* (or quadratic mean) described in the textbook. The variance is $s^2 = 122.4$ years^2 (rounded), and the standard deviation is 11.1 years (rounded). The value listed as "Mean Abs. Dev" is the mean absolute deviation described in the textbook. Also see that a histogram is displayed. There are other results that will be discussed later.

Descriptive statistics, including the measures of center and variation, can also be found by selecting **Data**, then **Descriptive Statistics**, but using the **Explore Data** feature is much better in the sense that more results are provided. Disadvantages of using Explore Data are (1) it is based on a single list of data values, and (2) the histogram uses default settings. If you prefer to use boxplots to compare two or more data sets, use **Data/Boxplot** instead of **Data/Explore Data**. If you prefer to generate a histogram using your own class width and starting point instead of the default settings, use **Data/Histogram** instead of **Data/Explore Data**.

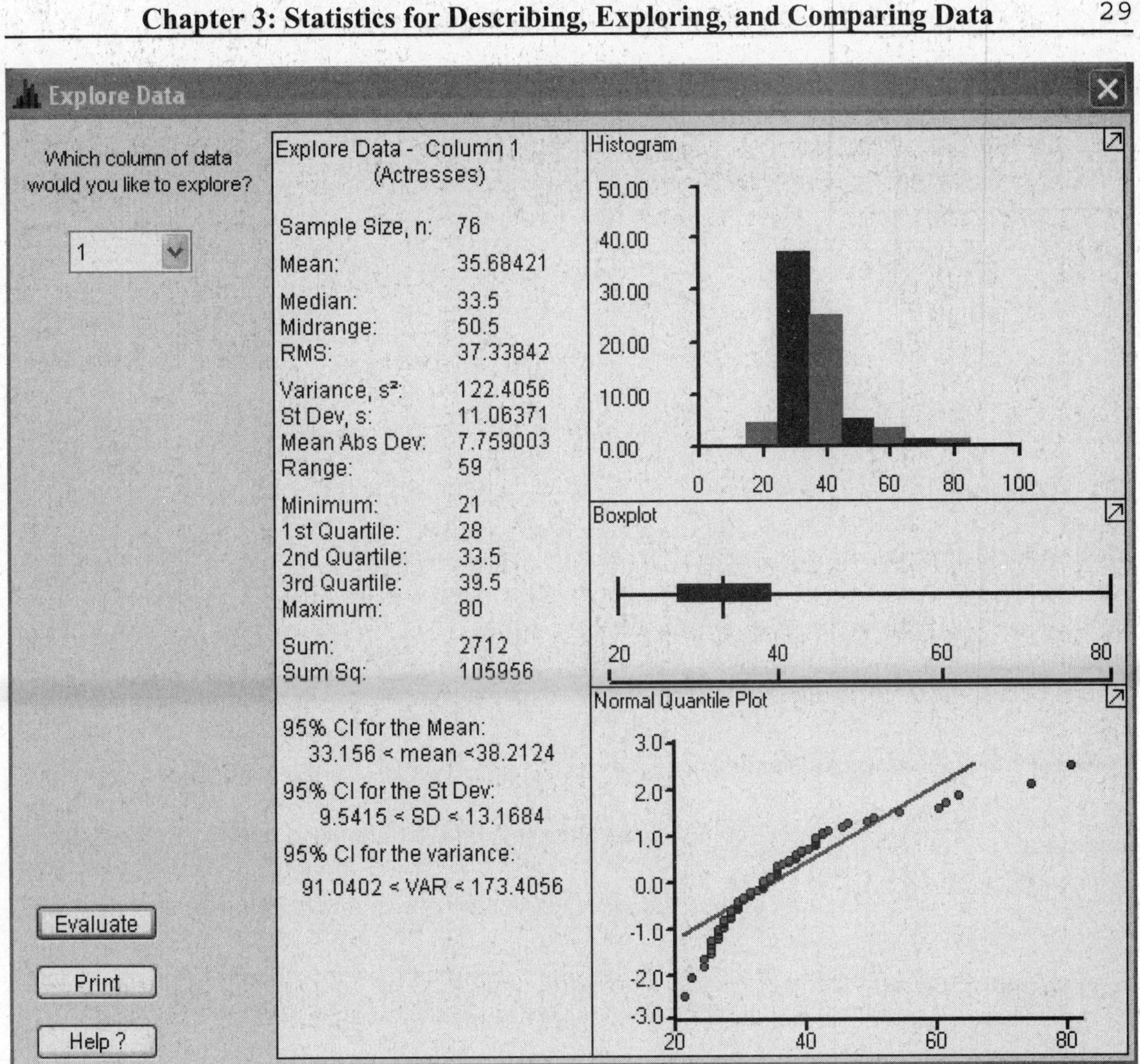

3-2 Quartiles and 5-Number Summary

The textbook includes the definition of a "5-number summary (minimum, 1st quartile, 2nd quartile, 3rd quartile, maximum), and that summary is included with the above STATDISK results. Here is the 5-number summary:

Minimum:	21 years
1st Quartile Q_1:	28 years
2nd Quartile Q_2:	33.5 years
3rd Qaurtile Q_3:	39.5 years
Maximum:	80 years

Important Note: The textbook states that there is not universal agreement on a single procedure for calculating quartiles, and different computer programs might yield different results. For example, if you use the ages of actresses listed earlier, you will get the results shown below. For this particular data set, STATDISK and the TI-83/84 Plus calculator agree, but they do not always agree.

	Q_1	Q_2	Q_3
STATDISK	28	33.5	**39.5**
Minitab	28	33.5	**39.75**
Excel	28	33.5	**39.25**
TI-83/84 Plus	28	33.5	**39.5**

3-3 Boxplots

The textbook describes the construction of boxplots. They are based on the 5-number summary consisting of the minimum, first quartile, second quartile, third quartile, and maximum. A boxplot is included among the results when STATDISK's Explore Data feature is used, but when using boxplots to compare two or more data sets, it is better to use the following procedure that allows you to generate two or more boxplots in the same display so that comparisons become easier.

Procedure for Generating a Boxplot

1. Enter or retrieve a set of sample data using one of these procedures:

    - **Manual entry of data:** Values can be entered in the Statdisk data window.
    - **Retrieve a data set from those included in Appendix B:** Click on the main menu item of **Datasets** and proceed to select one of the listed data sets.
    - **Retrieve a data set that you created:** Use **File/Open** as described in Section 1-4 of this manual/workbook.

2. Click on **Data** in the main menu bar at the top.

3. Click on **Boxplot**.

4. Select the column(s) to be used for the creation of one or more boxplots. The default is column 1, and it can be changed to any column number between 1 and 9. (If you don't want to include column 1, remove it by clicking on the box with the check mark next to column 1. Click a box to insert a check mark or to remove a check mark.)

5. Click on the **Plot** button.

One important advantage of boxplots is that they are very useful in comparing data sets. Shown below is the STATDISK display showing the two boxplots representing the ages of actresses and the ages of actors as given in the preceding lists. In the display shown below, the top boxplot depicts the ages of actresses and the bottom boxplot represents the ages of actors. (Recall that we are working with the ages of actresses and actors at the times that they won Academy Awards.) Because the two boxplots are constructed on the same scale, a comparison becomes easier. The boxplots suggest that Oscar-winning actresses tend to be younger than Oscar-winning actors. The top boxplot representing the ages of Oscar-winning actresses extends farther to the right, but this is the result of a single actress who was 80 when she received her Oscar. Apart from that high age of 80, the boxplots generally suggest that Oscar-winning actresses tend to be younger than Oscar-winning actors.

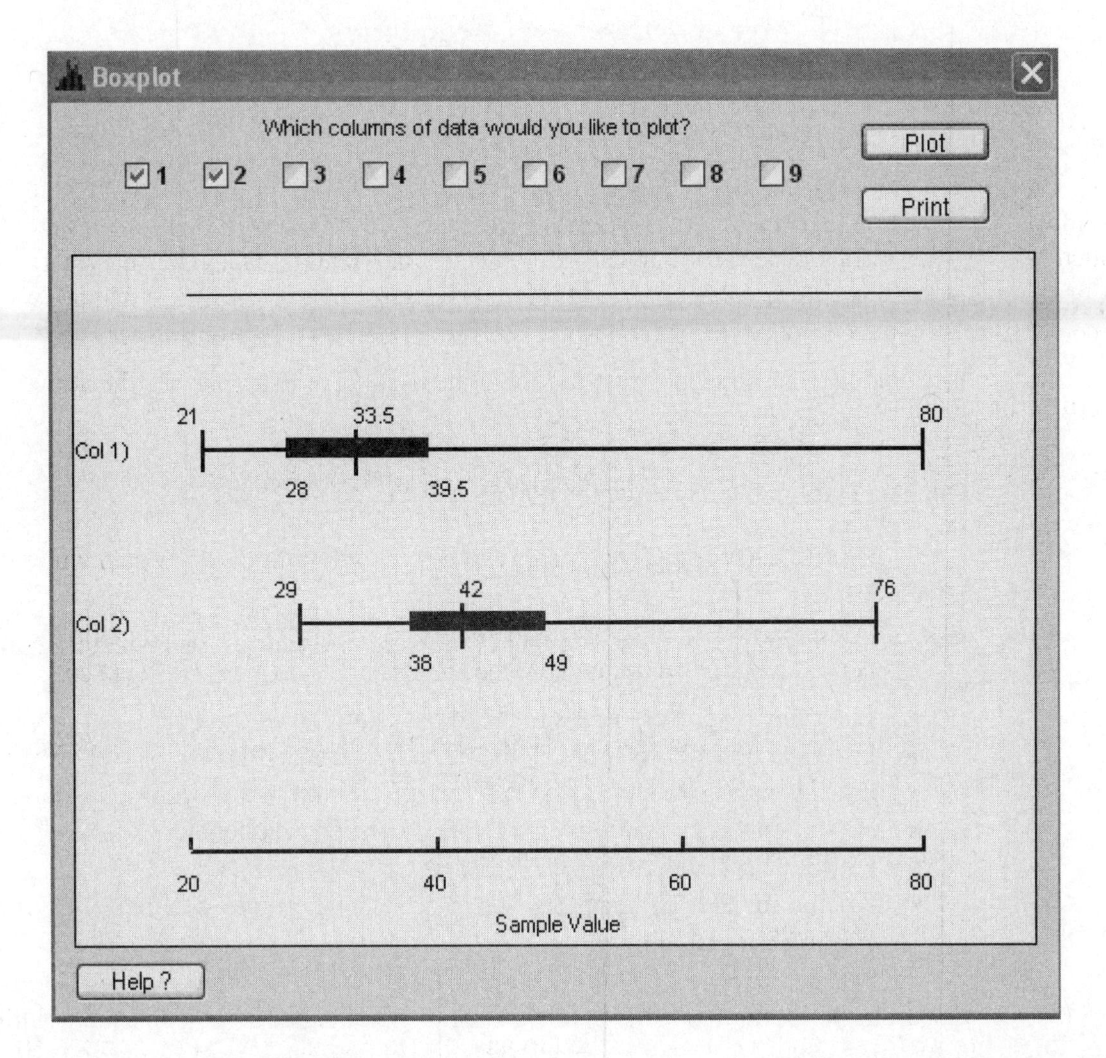

Important note: STATDISK generates boxplots based on the minimum, maximum, and three quartiles. STATDISK determines the values of the quartiles by following the same procedure described in the textbook, but other programs may use different procedures, so there may be some differences in boxplot results.

3-4 Statistics From a Frequency Distribution

If sample data are summarized in the form of a table representing a frequency distribution such as the one shown below, STATDISK can be used to obtain the important measures of center and variation. The basic idea is to use STATDISK's **Frequency Table Generator** to generate a list of sample values based on the table. Given the table below, for example, STATDISK can generate a list containing 28 values of 25.5 (the midpoint of the first class), 30 values of 35.5 (the midpoint of the second class), and so on. The result will be a list of 76 sample values that correspond to the frequency distribution.

Age of Actress	Frequency
21-30	28
31-40	30
41-50	12
51-60	2
61-70	2
71-80	2

Procedure for Obtaining Statistics From a Frequency Distribution

1. First identify the frequency distribution to be used. (For example, see the above table.)

2. Click on **Data** in the main menu bar at the top.

3. Click on **Frequency Table Generator**.

4. See the following STATDISK display showing the entries corresponding to the above frequency distribution.
   - Enter the lower class limits in the "Start" column as shown.
   - Enter the upper class limits in the "End" column as shown.
   - Enter the class frequencies in the "Freq." column as shown.
   - Be sure to select "Sample with Same Observed Frequencies."
   - For the output values, select "Equal to Class Midpoints."
   - For the number of decimals, select at least one more decimal place than is used for the class limits.

5. Click on the **Generate Data** button to get a list of sample values in the column at the right.

6. After the data have been generated and they appear in the column at the right, click on **Copy** and proceed to the main STATDISK data window where the list of values can be pasted. Once the list of sample values is copied to the data window, you can use **Explore Data** to obtain the descriptive statistics.

Caution: When using the above procedure, realize that you are not generating a list of sample values with the exact same characteristics as the original list of sample data. If the original sample values are not known, there is no way to recreate the original list from a frequency distribution table. Statistics calculated from the generated data are likely to differ somewhat from the statistics that would be calculated using the original list of sample data. For example, using the original list of ages of Oscar-winning actresses, the mean is found to be 35.68421 years (or 35.7 years when rounded), but using the generated values from the frequency distribution results in a mean of 35.76315 years (or 35.8 years when rounded).

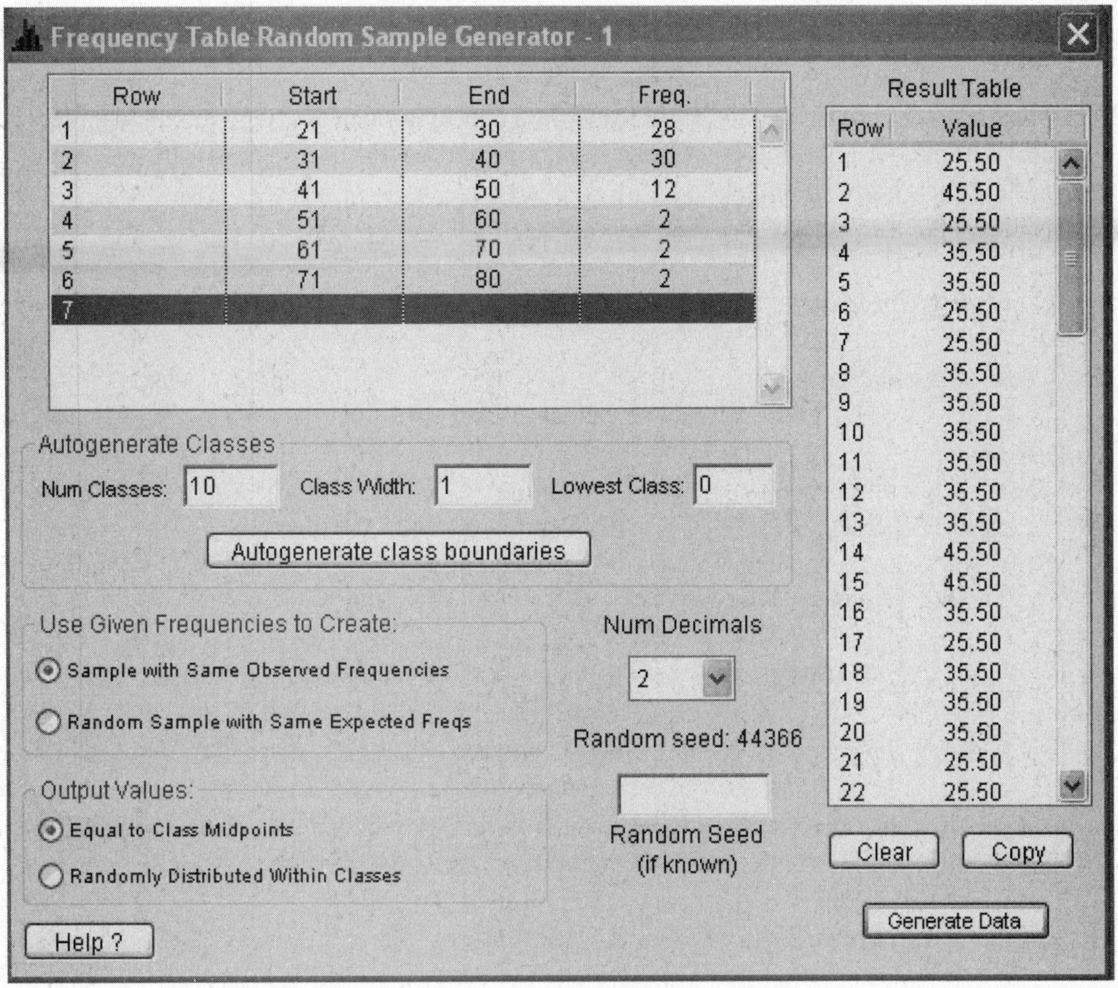

Frequency Table Random Sample Generator - 1

Row	Start	End	Freq.
1	21	30	28
2	31	40	30
3	41	50	12
4	51	60	2
5	61	70	2
6	71	80	2
7			

Autogenerate Classes
Num Classes: 10 Class Width: 1 Lowest Class: 0
Autogenerate class boundaries

Use Given Frequencies to Create:
Sample with Same Observed Frequencies
Random Sample with Same Expected Freqs

Num Decimals 2
Random seed: 44366
Random Seed (if known)

Output Values:
Equal to Class Midpoints
Randomly Distributed Within Classes

Help ?

Result Table

Row	Value
1	25.50
2	45.50
3	25.50
4	35.50
5	35.50
6	25.50
7	25.50
8	35.50
9	35.50
10	35.50
11	35.50
12	35.50
13	35.50
14	45.50
15	45.50
16	35.50
17	25.50
18	35.50
19	35.50
20	35.50
21	25.50
22	25.50

Clear Copy
Generate Data

CHAPTER 3 EXPERIMENTS: Statistics for Describing, Exploring, and Comparing Data

3–1. ***Describing Data*** Use the 40 BMI (body mass index) indices of women from the data set Fhealth (female health) and enter the results indicated below.

Center: Mean: _____ Median: _____

Variation: St. Dev.:_____ Range: _____

5-Number Summary: Min.:_____ Q_1:_____ Q_2:_____ Q_3:_____ Max.:_____

Outliers: ___________________

3–2. ***Describing Data*** Use the 40 BMI (body mass index) indices of men from the data set Mhealth (male health) and enter the results indicated below.

Center: Mean: _____ Median: _____

Variation: St. Dev.:_____ Range: _____

5-Number Summary: Min.:_____ Q_1:_____ Q_2:_____ Q_3:_____ Max.:_____

Outliers: ___________________

Compare the BMI indices of women (from Experiment 3-1) and men.

____

____

____

3–3. ***Describing Data*** Use the 40 pulse rates of women from the data set Fhealth (female health) and enter the results indicated below.

Center: Mean: _____ Median: _____

Variation: St. Dev.:_____ Range: _____

5-Number Summary: Min.:_____ Q_1:_____ Q_2:_____ Q_3:_____ Max.:_____

Outliers: ___________________

3–4. ***Describing Data*** Use the 40 pulse rates of men from the data set Mhealth (male health) and enter the results indicated below.

Center: Mean: _____ Median: _____

Variation: St. Dev.: _____ Range: _____

5-Number Summary: Min.: _____ Q_1: _____ Q_2: _____ Q_3: _____ Max.: _____

Outliers: ____________________

Compare the pulse rates of women (from Experiment 3-3) and men.

____

____

3–5. ***Boxplots*** Use the BMI indices of women (in data set Fhealth) and men (in data set Mhealth) and print their two boxplots together in the same display. (*Hint:* Because the two lists are in different data sets, use Copy/Paste to configure the two lists of values so that they are in the same data window.) Do the boxplots suggest any notable differences in the two sets of sample data?

____

____

3–6. ***Boxplots*** Use the pulse rates of women (in data set Fhealth) and men (in data set Mhealth) and print their two boxplots together in the same display. (*Hint:* Because the pulse rates of women and men are in different data sets, use Copy/Paste to configure the two lists of values so that they are in the same data window.) Do the boxplots suggest any notable differences in the two sets of sample data?

____

____

3–7. ***Tobacco/Alcohol Use in Children's Movies*** In "Tobacco and Alcohol Use in G–Rated Children's Animated Films," by Goldstein, Sobel and Newman (*Journal of the American Medical Association,* Vol. 281, No. 12), the lengths (in seconds) of scenes showing tobacco use and alcohol use were recorded for animated children's movies. Refer to the data in Appendix B from the textbook and find the mean and median for the tobacco times, then find the mean and median for the alcohol times. Does there appear to be a difference between those times? Which appears to be the larger problem: scenes showing tobacco use or scenes showing alcohol use?

____

____

3–8. ***Effect of Outlier*** Experiment 3–1 used the BMI measurements of 40 females. Change the first BMI value from 19.6 to the outlier of 196, then repeat Experiment 3-1. Enter the results below.

Center: Mean: _____ Median: _____

Variation: St. Dev.:_____ Range: _____

5-Number Summary: Min.:_____ Q_1:_____ Q_2:_____ Q_3:_____ Max.:_____

Outliers: ____________________

Based on a comparison of these results to those found in Experiment 3–1, how is the mean affected by the presence of an outlier?

____

How is the median affected by the presence of an outlier?

____

How is the standard deviation affected by the presence of an outlier?

____

3–9. ***Interpreting Dotplot*** Shown below is a dotplot of sample data. Identify the values represented in this graph, enter them in STATDISK, then find the indicated results.

3 4 5 6 7 8 9 10

Center: Mean: _____ Median: _____

Variation: St. Dev.:_____ Range: _____

5-Number Summary: Min.:_____ Q_1:_____ Q_2:_____ Q_3:_____ Max.:_____

Outliers: ____________________

3–10. ***Frequency Distribution*** Use the frequency distribution below to first generate a list of sample data, then find the indicated statistics and enter them here.

Mean: ________ Standard deviation: ________

Speed	Frequency
42–45	25
46–49	14
50–53	7
54–57	3
58–61	1

3–11. ***Frequency Distribution*** Use the frequency distribution below to first generate a list of sample data, then find the indicated statistics and enter them here.

Mean: ________ Standard deviation: ________

How does the mean compare to the value of 98.6°F, which is assumed to be the population mean by many people?

____

____

Temperature	Frequency
96.5–96.8	1
96.9–97.2	8
97.3–97.6	14
97.7–98.0	22
98.1–98.4	19
98.5–98.8	32
98.9–99.2	6
99.3–99.6	4

3–12. ***Comparing Data*** Open the data set COLA and use STATDISK to compare the weights of regular Coke and the weights of diet Coke. Obtain printouts of relevant results. What do you conclude? Can you explain any substantial difference?

____

____

3–13. ***Working with Your Own Data*** Through observation or experimentation, collect your own set of sample values. Obtain at least 40 values and try to select data from an interesting population. Use STATDISK to explore the data. Obtain printouts of relevant results. Describe the nature of the data. That is, what do the values represent? Describe important characteristics of the data set, and include printouts to support your observations.

____

____

____

____

3–14. ***Activities with STATDISK: Exploring Standard Deviation*** Explore the meaning of standard deviation by processing the following data sets. First enter the values in the data window as shown below.

Sample Editor

Enter your sample values in the columns below.
You can access these data directly from other modules.
For detailed instructions, click the help button.

Row	1	2	3	4	5	6	7	8	9
1	1	1	0.10	1000.1	10	10000	10000	10000	10000000
2	2	11	0.20	1000.2	11	10001	10100	10010	1000000
3	3	21	0.30	1000.3	12	10002	10200	10020	100000
4	4	31	0.40	1000.4	13	10003	10300	10030	10000
5	5	41	0.50	1000.5	14	10004	10400	10040	1000
6	6	51	0.60	1000.6	15	10005	10500	10050	100
7	7	61	0.70	1000.7	16	10006	10600	10060	10
8	8	71	0.80	1000.8	17	10007	10700	10070	1
9	9	81	0.90	1000.9	18	10008	10800	10080	0.1
10	10	91	1.00	1001.0	19	10009	10900	10090	0.01
11	11	101	1.10	1001.1	20	10010	11000	10100	0.001
12	12	111	1.20	1001.2	21	10011	11100	10110	0.0001
13	13	121	1.30	1001.3	22	10012	11200	10120	0.00001
14	14	131	1.40	1001.4	23	10013	11300	10130	0.000001
15	15	141	1.50	1001.5	24	10014	11400	10140	0.0000001

Process the data sets by using the Explore Data module. Complete the following table by recording the values from STATDISK, and use your calculator to find the coefficients of variation.

Data Set by Column	Standard Deviation	Mean	Coefficient of Variation
1			
2			
3			
4			
5			
6			
7			
8			
9			

Which three data sets have the same standard deviation?

___

Apart from having the same value of the standard deviation, describe a way in which the three data sets are related.

___

What do you notice about the values of the data in Column 9? What effect does this pattern have on the standard deviation in that set?

___

Do data sets with large values have large standard deviations, while those data sets with small values have small standard deviations? Why or why not?

___

What characteristic of a data set is most important in determining standard deviation?

___

The coefficient of variation measures standard deviation as a percent of the mean. Which coefficient of variation is the largest? Which is the smallest?

___

Considering all nine data sets, what general trend do you observe that explains the differences among the coefficients of variation? Focus on each standard deviation and corresponding mean.

___

___

3–15. ***Activities with STATDISK: Exploring Characteristics of Data Sets***

When a sample size is small, it is not easy to characterize the shape of a distribution. Refer to the nine data sets listed in Experiment 3-14. Before considering distributions of the nine data sets, study the appearance of large data sets by using STATDISK's Normal Generator and Uniform Generator.

Select **Data**, then select **Normal Generator**. Use the Normal Generator module four times, using the following. After generating each data set, use Copy/Paste to copy the data to columns of the STATDISK data window. (Use columns 1, 2, 3, 4.) Note that *the same random seed is used for all eight of the following generated data sets.*

n = 50; s = 75; 0 decimal places; use the random seed generated by STATDISK.
n = 100; s = 75; 0 decimal places; use the same random seed from the first sample.
n = 500; s = 75; 0 decimal places; use the same random seed from the first sample.
n = 1000; s = 75; 0 decimal places; use the same random seed from the first sample.

Select **Data**, then select **Uniform Generator**. Use the Uniform Generator module four times, using the following. After generating each data set, use Copy/Paste to copy the data to columns of the STATDISK data window. (Use columns 5, 6, 7, 8). Note that *these four samples use the same random seed generated in the first sample above.*

n = 50; s = 75; 0 decimal places; use the same random seed from the first sample.
n = 100; s = 75; 0 decimal places; use the same random seed from the first sample.
n = 500; s = 75; 0 decimal places; use the same random seed from the first sample.
n = 1000; s = 75; 0 decimal places; use the same random seed from the first sample.

Use STATDISK to generate a histogram for each of the eight data sets. Which histogram best portrays a bell-shaped distribution?

____

Which histogram best portrays a flat distribution?

____

Use STATDISK to generate and print all eight boxplots together. What do you observe about the results?

____

Considering all of the preceding results, which type of graph is best for identifying the shape of a distribution?

____

continued

Identify the shapes of the eight distributions.

Column 1: ______________________________

Column 2: ______________________________

Column 3: ______________________________

Column 4: ______________________________

Column 5: ______________________________

Column 6: ______________________________

Column 7: ______________________________

Column 8: ______________________________

Probabilities through Simulations

4-1 Simulation Methods

The textbook presents a variety of rules and methods for finding probabilities of different events. The probability chapter in the textbook focuses on traditional approaches to computing probability values. This chapter focuses instead on an alternative approach based on *simulations*.

> A **simulation** of a procedure is a process that behaves the same way as the procedure, so that similar results are produced.

Mathematician Stanislaw Ulam once studied the problem of finding the probability of winning a game of solitaire, but the theoretical computations involved were too complicated. Instead, Ulam took the approach of programming a computer to simulate or "play" solitaire hundreds of times. The ratio of wins to total games played is the approximate probability he sought. This same type of reasoning was used to solve important problems that arose during World War II. There was a need to determine how far neutrons would penetrate different materials, and the method of solution required that the computer make various random selections in much the same way that it can randomly select the outcome of the rolling of a pair of dice. This neutron diffusion project was named the Monte Carlo Project and we now refer to general methods of simulating experiments as *Monte Carlo methods*. Such methods are the focus of this chapter. The concept of simulation is quite easy to understand with simple examples.

- We could simulate the rolling of a die by using STATDISK to randomly generate whole numbers between 1 and 6 inclusive, provided that the computer selects from the numbers 1, 2, 3, 4, 5, and 6 in such a way that those outcomes are equally likely.
- We could simulate births by flipping a coin, where "heads" represents a baby girl and "tails" represents a baby boy. We could also simulate births by using STATDISK to randomly generate 1s (for baby girls) and 0s (for baby boys).

It is extremely important to construct a simulation so that it behaves just like the real procedure. See this following example and observe the right way and the wrong way of constructing the simulation.

EXAMPLE Describe a procedure for simulating the rolling of a pair of dice.

SOLUTION In the procedure of rolling a pair of dice, each of the two dice yields a number between 1 and 6 (inclusive) and those two numbers are then added. Any simulation should do exactly the same thing.

Right *way to simulate rolling two dice:* Randomly generate one number between 1 and 6, then randomly generate another number between 1 and 6, then add the two results.

Wrong *way to simulate rolling two dice:* Randomly generate numbers between 2 and 12. This procedure is similar to rolling dice in the sense that the results are always between 2 and 12, but these outcomes between 2 and 12 are equally likely. With real dice, the values between 2 and 12 are *not* equally likely. This simulation would yield terrible results.

4-2 STATDISK Simulation Tools

STATDISK includes several different tools that can be used for simulations. If you click on the main menu item of **Data**, you get a submenu that includes these items:

Normal Generator
Uniform Generator
Binomial Generator
Poisson Generator
Coins Generator
Dice Generator
Frequency Table Generator

Here are descriptions of these menu items available by clicking on **Data**:

- **Normal Generator**: Generates a sample of data randomly selected from a population having a normal distribution. You must enter the population mean and standard deviation, along with the number of values to be generated. (Normal distributions are described in the textbook. For now, consider a normal distribution to be a distribution that is bell–shaped.)

- **Uniform Generator:** This tool is particularly good for the *random generation of integers*. It generates numbers between a desired minimum value and maximum value. You can specify the number of decimal places, so enter 0 if you want only whole numbers. The generated values are "uniform" in the sense that all possible values have the same chance of being selected. For example, if you select a sample size of 500, a minimum of 1, a maximum of 6, and 0 decimal places, the results simulate the rolling of a single die 500 times, as shown in the following STATDISK display. (See also the Dice Generator described below.)

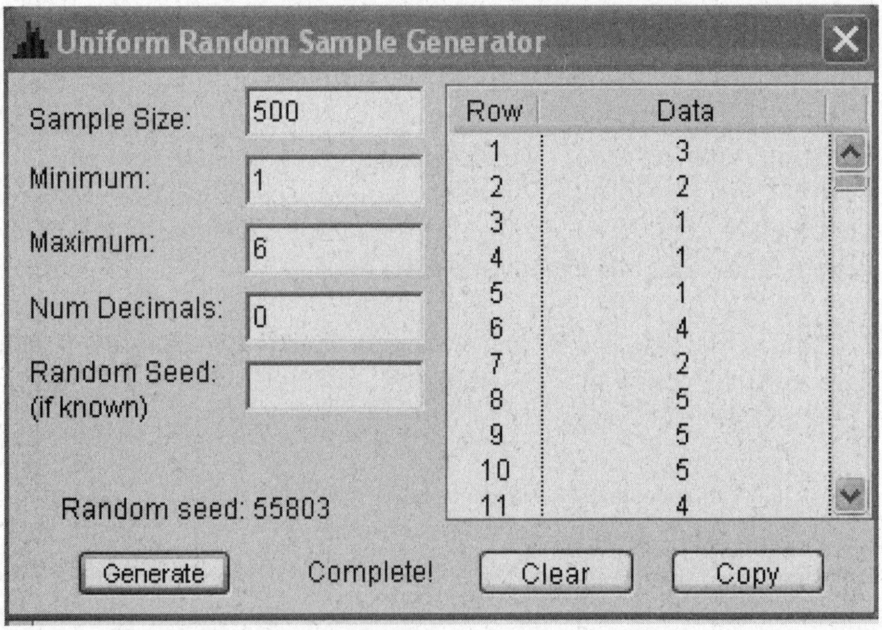

Uniform Random Sample Generator

Sample Size: 500
Minimum: 1
Maximum: 6
Num Decimals: 0
Random Seed: (if known)

Random seed: 55803

Row	Data
1	3
2	2
3	1
4	1
5	1
6	4
7	2
8	5
9	5
10	5
11	4

Generate Complete! Clear Copy

- **Binomial Generator**: Generates numbers of successes for a binomial probability distribution. You specify the number of values to be generated, the probability of success, and the number of trials in each case. Binomial probability distributions are discussed later in the textbook, so this item can be ignored for now.

- **Coins Generator**: This tool is particularly useful for those cases in which there are two possible outcomes (such as boy/girl) that are equally likely, as is the case with coin tosses. You select the number of generated values that you want, and you also select the number of coins to be tossed. The generated values are the numbers of heads that turn up.

- **Dice Generator:** Select the number of generated values that you want, and select the number of dice to be rolled. Also select the number of sides the dice have (use 6 for standard dice). The generated values are the totals of the dice.

- **Frequency Table Generator:** This feature can be used to generate sample data drawn from a population that can be described by a frequency distribution. Click on the main menu item of **Data** and then select **Frequency Table Generator** to get the window shown below. You can automatically generate the class boundaries. There are other options indicated in the window. Click **Generate Data** when you are ready. The generated data can be copied to the Statdisk data window where it can be used with other modules, such as Descriptive Statistics or Histogram.

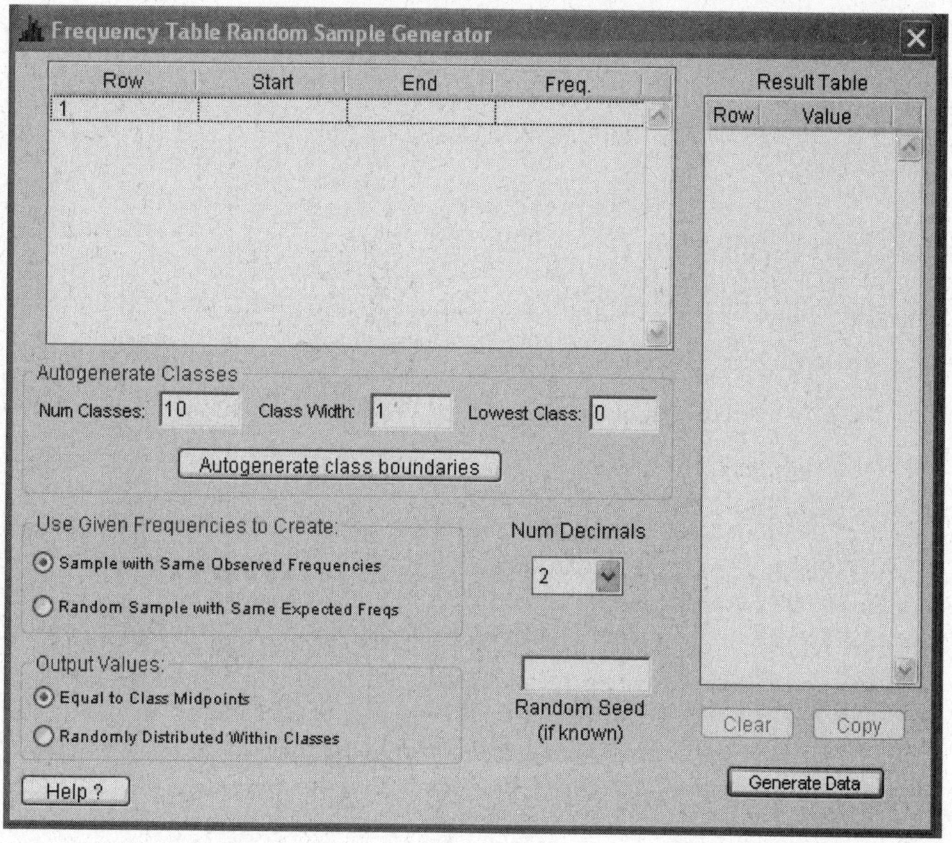

Random Seed

The above STATDISK tools include an option for entering a "random seed" if it is known. This entry will be usually left blank, but if you either record a seed that was used or if you enter a value for your own seed, you can duplicate results that were previously obtained. For example, an instructor might assign the generation of data with a particular random seed so that everyone in the class will get identical results. Most of the time, you will *not* enter a value for the random seed so that your results will be different each time. This makes life a bit more interesting.

4-3 Sorting Data

In some cases, it is very helpful to *sort* data. (That is, arrange the data in order.) In Section 2-5 of this manual/workbook we described the procedure for sorting data. This procedure is slightly modified to begin with data that have been generated:

Procedure for Sorting Data

1. After generating sample data using the tools described in Section 4-2 of this manual/workbook, use Copy/Paste to copy the data to the Statdisk data window.

2. With the data set listed in Statdisk data window, click the **Data tools** bar.

3. A window will appear with an option of **Sort data**.

4. Proceed to sort the column of generated values.

4-4 Simulation Examples

We will now illustrate the preceding STATDISK features by describing specific simulations.

Simulation 1: Generating 50 births (boys/girls)

To simulate 50 births with the assumption that boys and girls are equally likely, use either of the following:

- Use STATDISK's **Uniform Generator** (see Section 4-2) to generate 50 integers between 0 and 1. Be sure to enter 0 for the number of decimal places. If you arrange the results in order, it is very easy to count the number of 0s (or boys) and the number of 1s (or girls). See Section 4-3 of this manual/workbook for the procedure for sorting data.

- Use STATDISK's **Coins Generator** (see Section 4-2). Enter 50 for the sample size and enter 1 for the number of coins. Again, it is very easy to count the number of 0s (or boys) and the number of 1s (or girls) if the data are sorted, as described in Section 4-3 of this manual/workbook.

Simulation 2: Rolling a single die 60 times

To simulate 60 rolls of a single die, use either of these approaches:

- Use STATDISK's **Uniform Generator** (see Section 4-2) to generate 60 integers between 1 and 6. (Enter 60 for the sample size and be sure to enter 0 for the number of decimal places.) Again, arranging them in order makes it easy to count the number of 1s, 2s, and so on.

- Use STATDISK's **Dice Generator**. Enter 60 for the sample size, enter 1 for the number of dice, and enter 6 for the number of sides.

Simulation 3: Generating 25 birth dates

Instead of generating 25 results such as "January 1," or "November 27," randomly generate 25 integers between 1 and 365 inclusive. (We are ignoring leap years). Use STATDISK's **Uniform Generator** and enter 25 for the sample size. Also enter 1 for the minimum, 365 for the maximum, and be sure to enter 0 for the number of decimal places (so that only integers are generated). See the sample STATDISK display shown below. The first generated value of 341 corresponds to December 7, because the 341st day in a year is December 7.

Even though the display shows only the first 11 of the 25 birth dates, we can examine the complete list to determine whether two values occur twice. You can examine all 25 entries by scrolling, but if you sort the simulated birth dates by copying the list of data to the STATDISK data window and using the sort feature (by clicking on the **Data tools** button), it becomes much easier to scan the sorted list and determine whether there are two birth dates that are the same. If there are two birth dates that are the same, they will show up as *consecutive* equal values in the sorted list.

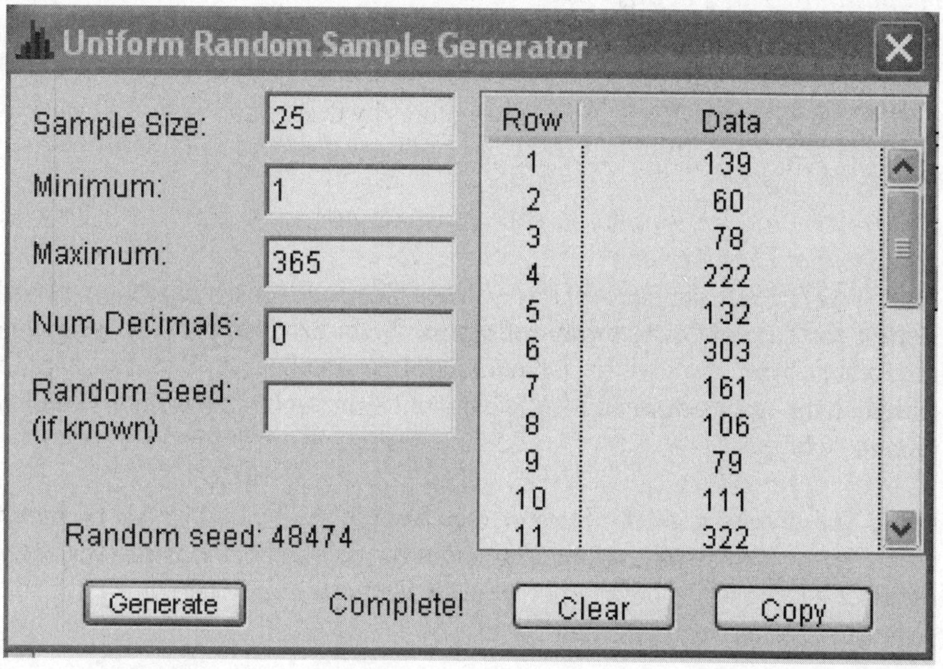

Uniform Random Sample Generator

Sample Size: 25
Minimum: 1
Maximum: 365
Num Decimals: 0
Random Seed: (if known)

Random seed: 48474

Row	Data
1	139
2	60
3	78
4	222
5	132
6	303
7	161
8	106
9	79
10	111
11	322

Generate Complete! Clear Copy

CHAPTER 4 EXPERIMENTS: Probabilities through Simulations

4-1. ***Birth Simulation*** Use STATDISK to simulate 500 births, where each birth results in a boy or girl. Sort the results, count the number of girls, and enter that value here:_______

Based on that result, estimate the probability of getting a girl when a baby is born. Enter the estimated probability here: __________

The preceding estimated probability is likely to be different from 0.5. Does this suggest that the computer's random number generator is defective? Why or why not?

____

____

4-2. ***Dice Simulation*** Use STATDISK to simulate 1000 rolls of a pair of dice. Sort the results, then find the number of times that the total was exactly 7. Enter that value here:_____

Based on that result, estimate the probability of getting a 7 when two dice are rolled. Enter the estimated probability here:__________

How does this estimated probability compare to the theoretical probability of 0.167?

____

____

4-3. ***Probability of Exactly 11 Girls***

a. Use STATDISK to simulate 20 births. Does the result consist of exactly 11 girls?____
b. Repeat part (a) nine more times and record the result from part (a) along with the other nine results here: ___ ___ ___ ___ ___ ___ ___ ___ ___ ___
c. Based on the results from part (b), what is the estimated probability of getting exactly 11 girls in 20 births? __________

4-4. ***Probability of at Least 11 Girls***

a. Use STATDISK to simulate 20 births. Does the result consist of at least 11 girls?____
b. Repeat part (a) nine more times and record the result from part (a) along with the other nine results here: ___ ___ ___ ___ ___ ___ ___ ___ ___ ___
c. Based on the results from part (b), what is the estimated probability of getting at least 11 girls in 20 births? __________

4-5. ***Guessing Simulation*** Use STATDISK to conduct a simulation that can be used to estimate the probability of getting at least six correct responses when random guesses are made for all 10 true/false questions in a quiz. Enter the estimated probability here: _____
Describe the simulation process that was used.

____

4-6. ***Guessing Simulation*** Use STATDISK to conduct a simulation that can be used to estimate the probability of getting at least six correct responses when random guesses are made for all 10 multiple choice questions in a quiz. Each question has five possible answers (a, b, c, d, e) and only one of them is correct. Enter the estimated probability here: ________ Describe the simulation process that was used.

____

____

4-7. ***Probability of at Least 55 Girls*** Use STATDISK to conduct a simulation for estimating the probability of getting at least 55 girls in 100 births. Enter the estimated probability here:__________ Describe the procedure used to obtain the estimated probability.

____

In testing a gender-selection method, assume that the Biogene Technology Corporation conducted an experiment with 100 couples who were treated, and that the 100 births included at least 55 girls. What should you conclude about the effectiveness of the treatment?

____

4-8. ***Probability of at Least 65 Girls*** Use STATDISK to conduct a simulation for estimating the probability of getting at least 65 girls in 100 births. Enter the estimated probability here:__________

Describe the procedure used to obtain the estimated probability.

____

In testing a gender-selection method, if the Biogene Technology Corporation conducted an experiment with 100 couples who were treated, and the 100 births included at least 65 girls, what should you conclude about the effectiveness of the treatment?

____

4–9. ***Simulating Families of Five Children*** Develop a simulation for finding the probability of getting at least two girls in a family of five children. Simulate 100 families. Describe the simulation, then estimate the probability based on its results.

____

____

4–10. ***Simulating Three Dice*** Simulate the rolling of the three dice 100 times, then use it to estimate the probability of getting a total of 10 when three dice are rolled. Enter the probability here: __________

4–11. ***Simulating Left–Handedness*** Ten percent of us are left-handed. In a study of dexterity, people are randomly selected in groups of five. Develop a simulation for finding the probability of getting at least one left-handed person in a group of five. Simulate 100 groups of five. How does the probability compare to the correct result of 0.410, which can be found by using the probability rules in the Probability chapter in the textbook?

____

____

4–12. ***Simulating Hybridization*** When Mendel conducted his famous hybridization experiments, he used peas with green pods and yellow pods. One experiment involved crossing peas in such a way that 25% of the offspring peas were expected to have yellow pods. Develop a simulation for finding the probability that when two offspring peas are produced, at least one of them has yellow pods. Generate 100 pairs of offspring. How does the result compare to the correct probability of 7/16, which can be found by using the probability rules in the Probability chapter in the textbook?

____

____

4-13. ***Roulette*** Simulate the spinning of a roulette wheel 500 times by randomly generating 500 integers between 0 and 37. (American roulette wheels have slots numbered 0 through 36, plus another slot numbered 00. Consider a STATDISK-generated outcome of 37 to be 00 on the roulette wheel.) Arrange the results in ascending order. Assuming that you bet on the number 7 every time, how many times did you win? Based on the results, what is the estimated probability of winning if you bet on a single number?

Number of wins:______ *P*(win) = _____

In order to make a profit, your number of wins in 500 spins must be at least 14. Would you have made a profit?_______

4-14. ***Birthdays*** Simulate a class of 25 birth dates by randomly generating 25 integers between 1 and 365. (We will ignore leap years.) Arrange the birth dates in ascending order, then examine the list to determine whether at least two birth dates are the same. (This is easy to do, because any two equal integers must be next to each other.)

Generated "birth dates:" ___ ___ ___ ___ ___ ___ ___ ___ ___ ___ ___ ___ ___

___ ___ ___ ___ ___ ___ ___ ___ ___ ___ ___ ___

Are at least two of the "birth dates" the same? _________

4-15. ***Birthdays*** Repeat the preceding experiment nine additional times and record all ten of the yes/no responses here:

____ ____ ____ ____ ____ ____ ____ ____ ____ ____

Based on these results, what is the probability of getting at least two birth dates that are the same (when a class of 25 students is randomly selected)? __________

4-16. ***Birthdays*** Repeat Experiments 4-14 and 4-15 for 50 people instead of 25. Based on the results, what is the estimated probability of getting at least two birth dates that are the same (when a class of 50 students is randomly selected)? __________

4-17. ***Birthdays*** Repeat Experiments 4-14 and 4-15 for 100 people instead of 25. Based on the results, what is the estimated probability of getting at least two birth dates that are the same (when a class of 100 students is randomly selected)? __________

4-18. ***Normally Distributed Heights*** Simulate 1000 heights of adult women. (Adult women have normally distributed heights with a mean of 63.6 in. and a standard deviation of 2.5 in.) Arrange the data in ascending order, then examine the results and estimate the probability of a randomly selected woman having a height between 64.5 in. and 72 in. (Those were the height restrictions for women to fit into Russian Soyuz spacecraft when NASA and Russia ran joint missions.) Enter the estimated probability here: __________

4–19. ***Effectiveness of Drug*** It has been found that when someone tries to stop smoking under certain circumstances, the success rate is 20%. A new nicotine substitute drug has been designed to help those who wish to stop smoking. In a trial of 50 smokers who use the drug while trying to stop, it was found that 12 successfully stopped. The manufacturer argues that the 12 successes are better than the 10 that would be expected without the drug, so the drug is effective. Conduct a simulation of 50 smokers trying to stop, and assume that the drug has no effect, so the success rate continues to be 20%. Repeat the simulation several times and determine whether 12 successes could easily occur with an ineffective drug. What do you conclude about the effectiveness of the drug?

____

____

4-20. ***Normally Distributed IQ Scores*** IQ scores are normally distributed with a mean of 100 and a standard deviation of 15. Generate a normally distributed sample of 800 IQ scores by using the given mean and standard deviation. Sort the results (arrange them in ascending order).

a. Examine the sorted results to estimate the probability of randomly selecting someone with an IQ score between 90 and 110 inclusive. Enter the result here.______

b. Examine the sorted results to estimate the probability of randomly selecting someone with an IQ score greater than 115. __________

(continued)

c. Examine the sorted results to estimate the probability of randomly selecting someone with an IQ score less than 120. ________

d. Repeat part (a) of this experiment nine more times and list all ten probabilities here.

_____ _____ _____ _____ _____ _____ _____ _____ _____ _____

e. Examine the ten probabilities obtained above and comment on the *consistency* of the results.

____

____

f. How might we modify this experiment so that the results can become more consistent?

____

____

g. If the results appear to be very consistent, what does that imply about any individual sample result?

____

____

4-21. ***Law of Large Numbers*** In this experiment we test the Law of Large Numbers, which states that "as an experiment is repeated again and again, the empirical probability of success tends to approach the actual probability." We will use a simulation of a single die, and we will consider a success to be the outcome of a 1. [Based on the classical definition of probability, we know that $P(1) = 1/6 = 0.167$.]

a. Simulate 5 trials by generating 5 integers between 1 and 6. Count the number of 6s that occurred and divide that number by 5 to get the empirical probability.
Based on 5 trials, $P(1) =$ _____.

b. Repeat part (a) for 25 trials. Based on 25 trials, $P(1) =$ _____.

c. Repeat part (a) for 50 trials. Based on 50 trials, $P(1) =$ _____.

d. Repeat part (a) for 500 trials. Based on 500 trials, $P(1) =$ _____.

e. Repeat part (a) for 1000 trials. Based on 1000 trials, $P(1) =$ _____.

f. In your own words, generalize these results in a restatement of the Law of Large Numbers.

____

4-22. ***Sticky Probability Problem*** Consider the following statement taken from an exercise in the textbook. (It is noted that this is possibly the most difficult exercise in the entire book, but the use of simulation tools makes the solution much easier.)

Two points along a straight stick are randomly selected. The stick is then broken at these two points. Find the probability that the three pieces can be arranged to form a triangle.

While a theoretical solution is possible, it is very difficult. Instead, we will use a computer simulation. The length of the stick is irrelevant, so assume it's one unit long and its length is measured from 0 at one end to 1 at the other end. Use STATDISK to randomly select the two break points with the random generation of two numbers from a uniform distribution with a minimum of 0, a maximum of 1, and 4 decimal places. Plot the break points on the "stick" below.

0____1

A triangle can be formed if the longest segment is less than 0.5, so enter the lengths of the three pieces here: __________ __________ __________

Can a triangle be formed?

Now repeat this process nine more times and summarize all of the results below.

Trial	Break Points		Triangle formed?
1			
2			
3			
4			
5			
6			
7			
8			
9			
10			

Based on the ten trials, determine the estimated probability that a triangle can be formed and enter that estimated probability here: __________
This estimate gets better with more trials.

Probability Distributions

5-1 Exploring Probability Distributions

5-2 Binomial Distributions

5-3 Poisson Distributions

5-4 Cumulative Probabilities

5-1 Exploring Probability Distributions

The "Probability Distributions" chapter in the textbook (Chapter 5 in *Elementary Statistics*, 10th edition) introduces the important concept of a probability distribution, and that chapter includes only *discrete* probability distributions. These important definitions are included:

Definitions

A **random variable** is a variable (typically represented by x) that has a single numerical value, determined by chance, for each outcome of a procedure.

A **probability distribution** is a graph, table, or formula that gives the probability for each value of the random variable.

When working with a probability distribution, we should consider the same important characteristics introduced in Chapter 2:

1. **Center:** Measure of center, which is a representative or average value that gives us an indication of where the middle of the data set is located

2. **Variation:** A measure of the amount that the values vary among themselves

3. **Distribution:** The nature or shape of the distribution of the data, such as bell-shaped, uniform, or skewed

4. **Outliers:** Sample values that are very far away from the vast majority of the other sample values

5. **Time:** Changing characteristics of the data over time

Section 5-1 of this manual/workbook addresses the above important characteristics for probability distributions. The characteristics of center and variation are addressed with formulas for finding the mean, standard deviation, and variance of a probability distribution. The characteristic of distribution is addressed through the graph of a probability histogram.

Although STATDISK is not designed to deal directly with a probability distribution, it can often be used. Let's consider the following typical exercise which requires that you first determine whether a probability distribution is defined by the given information and, if so, find the mean and standard deviation. If you examine the data in the table, you can verify that a probability distribution is defined because the two key requirements are satisfied: (1) The sum of the probabilities is equal to 1; (2) each individual probability is between 0 and 1 inclusive.

> ***Prior Sentences*** When randomly selecting a jail inmate convicted of DWI (driving while intoxicated), the probability distribution for the number x of prior DWI sentences is as described in the table on the following page (based on data from the U.S. Department of Justice).

x	$P(x)$
0	0.512
1	0.301
2	0.132
3	0.055

Having determined that the above table does define a probability distribution, let's now see how we can use STATDISK to find the mean and standard deviation. The basic approach is to use STATDISK's "Frequency Table Generator" to construct a table of actual values with the same distribution given in the table.

STATDISK Procedure for Working with a Probability Distribution

1. Click on **Data**.
2. Select the menu item of **Frequency Table Generator**.
3. Enter class limits and frequencies that correspond to the probability distribution. Shown below are the entries corresponding to the probability distribution given in the above table. See the first class where the value of 0 is represented by the class limits of -0.5 to 0.5 and a frequency of 512 (based on a probability of 0.512).

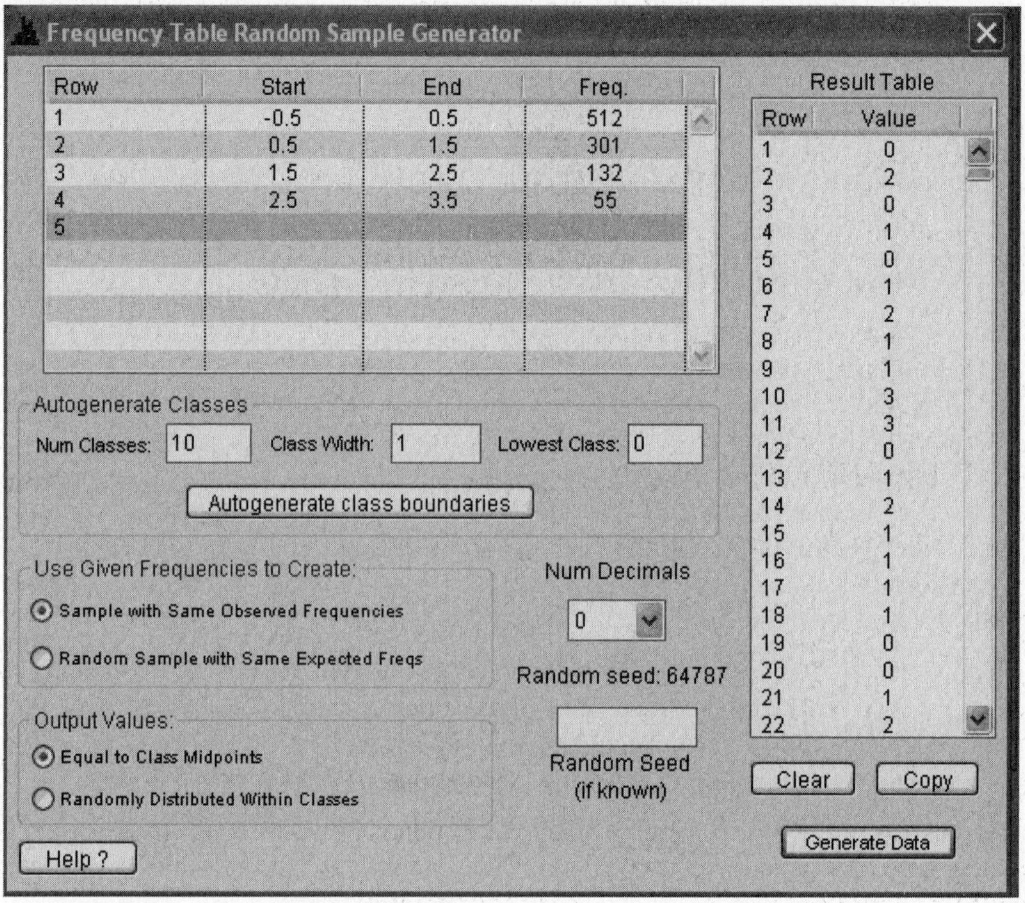

Frequency Table Random Sample Generator

Row	Start	End	Freq.
1	-0.5	0.5	512
2	0.5	1.5	301
3	1.5	2.5	132
4	2.5	3.5	55
5			

Result Table

Row	Value
1	0
2	2
3	0
4	1
5	0
6	1
7	2
8	1
9	1
10	3
11	3
12	0
13	1
14	2
15	1
16	1
17	1
18	1
19	0
20	0
21	1
22	2

Autogenerate Classes

Num Classes: 10 Class Width: 1 Lowest Class: 0

Autogenerate class boundaries

Use Given Frequencies to Create:

Sample with Same Observed Frequencies

Random Sample with Same Expected Freqs

Num Decimals: 0

Random seed: 64787

Output Values:

Equal to Class Midpoints

Randomly Distributed Within Classes

Random Seed (if known)

Clear Copy

Generate Data

Help ?

Note these settings in the preceding window:

- The number of decimals in the generated values is 0.
- Select "Samples with Same Observed Frequencies."
- Select output values "Equal to Class Midpoints."

4. Click on the **Generate Data** bar.
5. STATDISK will proceed to generate a set of data corresponding to the probability distribution.
6. You can now use Copy/Paste to copy the data to the Statdisk Data Window where you can find the mean and standard deviation or you can construct a histogram. There's one correction: If you copy the data to the Descriptive Statistics module, the computed standard deviation and variance could be off a little, because the calculation assumes *sample* data, whereas we should consider the data to be a *population*. If the sample size is large, the discrepancies will be small. For the data from the table reproduced above, the actual standard deviation is $\sigma = 0.88944$, but STATDISK yields $s = 0.88988$. The value of the mean will be correct. Here is a STATDISK histogram that shows the shape of the probability distribution:

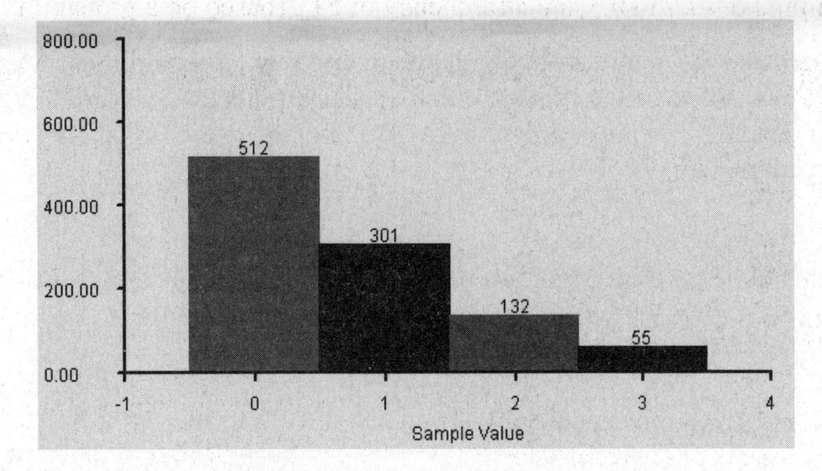

5-2 Binomial Distributions

In the textbook we define a **binomial distribution** to be a probability distribution that meets all of the following requirements:

1. The experiment must have a fixed number of trials.
2. The trials must be independent. (The outcome of any individual trial doesn't affect the probabilities in the other trials.)
3. Each trial must have all outcomes classified into two categories.
4. The probabilities must remain constant for each trial.

We also introduced notation with S and F denoting success and failure for the two possible categories of all outcomes. Also, p and q denote the probabilities of S and F, respectively, so that $P(S) = p$ and $P(F) = q$. We also use the following symbols.

n denotes the fixed number of trials

x denotes a specific number of successes in n trials so that x can be any whole number between 0 and n, inclusive.

p denotes the probability of success in *one* of the n trials.

q denotes the probability of failure in *one* of the n trials.

$P(x)$ denotes the probability of getting exactly x successes among the n trials.

The section of "Binomial Probability Distributions" in the textbook describes three methods for determining probabilities in binomial experiments. Method 1 uses the binomial probability formula:

$$P(x) = \frac{n!}{(n-x)!x!} p^x q^{n-x}$$

Method 2 requires use of Table A-1, the table of binomial probabilities. Method 3 requires computer usage. We noted in the textbook that if a computer and software are available, this third method of finding binomial probabilities is fast and easy, as shown in the following STATDISK procedure.

STATDISK Procedure for Finding Probabilities with a Binomial Distribution

1. Click on **Analysis** from the main menu.

2. Select **Binomial Probabilities**.

3. You will now see a dialog box, so make these entries:

   -Enter the number of trials n.
   -Enter the probability of success p.

4. Click on **Evaluate**.

Consider this example:

> **Analysis of Multiple Choice Answers** Use the binomial probability formula to find the probability of getting exactly 3 correct answers when random guesses are made for 4 multiple choice questions. That is, find the value $P(3)$ given that $n = 4$, $x = 3$, $p = 0.2$, and $q = 0.8$.

With STATDISK, click on **Analysis**, then select **Binomial Probabilities**, and proceed to enter 4 for the number of trials and 0.2 for the probability. Here is the STATDISK display:

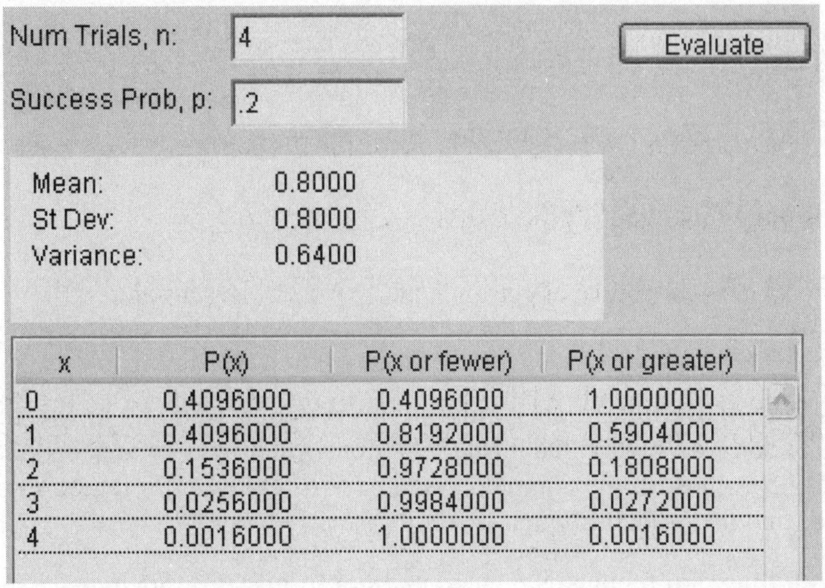

Num Trials, n: 4

Success Prob, p: .2

Evaluate

Mean: 0.8000
St Dev: 0.8000
Variance: 0.6400

x	P(x)	P(x or fewer)	P(x or greater)
0	0.4096000	0.4096000	1.0000000
1	0.4096000	0.8192000	0.5904000
2	0.1536000	0.9728000	0.1808000
3	0.0256000	0.9984000	0.0272000
4	0.0016000	1.0000000	0.0016000

From this STATDISK display, we can see that $P(3) = 0.02560$. Note that the display includes values of the mean, standard deviation, and variance. Also, STATDISK includes cumulative probabilities along with probabilities for the individual values of x. From the above display we can easily find probabilities such as these:

- The probability of 2 or fewer correct responses is 0.97280.
- The probability of 3 or more correct responses is 0.02720.

The table of binomial probabilities (Table A-1) in Appendix B of the textbook includes limited values of n and p, but STATDISK is so much more flexible in the values of n and p that can be used.

5-3 Poisson Distributions

See the textbook for a discussion of the Poisson distribution, where we see that it is a discrete probability distribution that applies to occurrences of some event *over a specified interval*. The random variable x is the number of occurrences of the event in an interval, such as time, distance, area, volume, or some similar unit. The probability of the event occurring x times over an interval is given by this formula:

$$P(x) = \frac{\mu^x \cdot e^{-\mu}}{x!} \quad \text{where } e \approx 2.71828$$

We also noted that the Poisson distribution is sometimes used to approximate the binomial distribution when $n \geq 100$ and $np \leq 10$; in such cases, we use $\mu = np$. If using STATDISK, the Poisson approximation to the binomial distribution isn't used as often, because we can easily find binomial probabilities for a wide range of values for n and p.

STATDISK Procedure for Finding Probabilities for a Poisson Distribution

1. Determine the value of the mean μ.

2. Click on **Analysis** from the main menu.

3. Select **Poisson Probabilities.**

4. Enter the value of the mean μ, then click on the **Evaluate** button.

Consider this example:

> **World War II Bombs** In analyzing hits by V-1 buzz bombs in World War II, South London was subdivided into 576 regions, each with an area of 0.25 km^2. A total of 535 bombs hit the combined area of 576 regions. If a region is randomly selected, find the probability that it was hit exactly twice.

With 535 bombs hitting 576 regions, the mean number of hits is 535/576 = 0.929. Having found the required mean μ, we can now proceed to use STATDISK. Click on **Analysis**, select **Poisson Probabilities,** and enter 0.929 for the mean. The result will be as shown below.

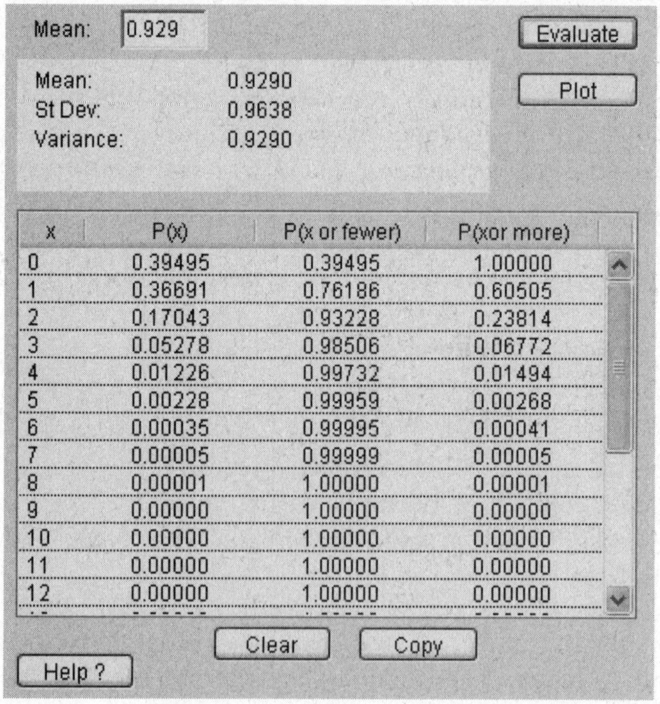

Mean: 0.929 Evaluate

Mean: 0.9290
St Dev: 0.9638
Variance: 0.9290

Plot

x	P(x)	P(x or fewer)	P(xor more)
0	0.39495	0.39495	1.00000
1	0.36691	0.76186	0.60505
2	0.17043	0.93228	0.23814
3	0.05278	0.98506	0.06772
4	0.01226	0.99732	0.01494
5	0.00228	0.99959	0.00268
6	0.00035	0.99995	0.00041
7	0.00005	0.99999	0.00005
8	0.00001	1.00000	0.00001
9	0.00000	1.00000	0.00000
10	0.00000	1.00000	0.00000
11	0.00000	1.00000	0.00000
12	0.00000	1.00000	0.00000

Clear Copy

Help ?

Note that display includes values for the mean, standard deviation, and variance. The probabilities and cumulative probabilities are listed in lower portion of the window. For example, $P(2)$ = 0.17043, which is the probability that a region would be hit two times. The probability of a region being hit 0, 1, or 2 times is 0.93228. The probability of x having a value of 2 or more is 0.23814, which is the probability of a region being hit at least twice.

5-4 Cumulative Probabilities

The main objective of this section is to reinforce the point that cumulative probabilities are often critically important. By *cumulative* probability, we mean the probability that the random variable x has a range of values instead of a single value. Here are typical examples:

- Find the probability of getting *at least* 13 girls in 14 births.
- Find the probability of getting *fewer than* 60 correct answers in 100 guesses to true/false questions.
- Find the probability of *more than* 5 wins when roulette is played 200 times.

As an example, consider a test of the MicroSort gender selection technique, with the result that there were 13 girls among 14 babies. Is this result unusual? Does this result really suggest that the technique is effective, or could it be that there were 13 girls among 14 babies just by chance? In answering this question, the relevant probability is the *cumulative* probability of getting 13 or more girls, not the probability of getting exactly 13 girls. The textbook supports this point with an example, reproduced here because it is so important:

> *Suppose you were flipping a coin to determine whether it favors heads, and suppose 1000 tosses resulted in 501 heads. This is not evidence that the coin favors heads, because it is very easy to get a result like 501 heads in 1000 tosses just by chance. Yet, the probability of getting exactly 501 heads in 1000 tosses is actually quite small: 0.0252. This low probability reflects the fact that with 1000 tosses, any specific number of heads will have a very low probability. However, we do not consider 501 heads among 1000 tosses to be unusual, because the probability of getting at least 501 heads is high: 0.487.*

The textbook notes that the principle used in the above example can be generalized as follows:

Using Probabilities to Determine When Results are Unusual

- x successes among n trials is unusually *high* if $P(x$ or more) is very small (such as 0.05 or less).

- x successes among n trials is unusually *low* if $P(x$ or fewer) is very small (such as 0.05 or less).

Cumulative probabilities therefore play a critical role in identifying results that are considered to be *unusual*. Later chapters focus on this important concept. If you examine the STATDISK displayed results for binomial probabilities and Poisson probabilities, you can see that in addition to obtaining probabilities for specific numbers of successes, it is very easy to obtain *cumulative* probabilities.

CHAPTER 5 EXPERIMENTS: Probability Distributions

5-1. Use STATDISK with the procedure described in Section 5-1 of this manual/workbook to find the mean and standard deviation of the frequency distribution given below. Also provide a printed display of a histogram of the probability distribution.

Mean: ________

St. Dev.: ________

Gender Selection *In a study of the MicroSort gender selection method, couples in a control group are not given a treatment and they each have three children. The probability distribution for the number of girls is given in the accompanying table.*

x	$P(x)$
0	0.125
1	0.375
2	0.375
3	0.125

5-2. Use STATDISK with the procedure described in Section 5-1 of this manual/workbook to find the mean and standard deviation of the frequency distribution given below. Also provide a printed display of a histogram of the probability distribution.

Mean: ________

St. Dev.: ________

Life Insurance *The Telektronic Company provides life insurance policies for its top four executives, and the random variable x is the number of those employees who live through the next year.*

x	$P(x)$
0	0.0000
1	0.0001
2	0.0006
3	0.0387
4	0.9606

5-3. Use STATDISK with the procedure described in Section 5-1 of this manual/workbook to find the mean and standard deviation of the frequency distribution given below. Also provide a printed display of a histogram of the probability distribution.

Mean: ________

St. Dev.: ________

Prior Sentences *When randomly selecting a jail inmate convicted of DWI (driving while intoxicated), the probability distribution for the number x of prior DWI sentences is as described in the accompanying table (based on data from the U.S. Department of Justice).*

x	$P(x)$
0	0.512
1	0.301
2	0.132
3	0.055

5-4. ***Binomial Probabilities*** Use STATDISK to find the binomial probabilities corresponding to $n = 4$ and $p = 0.05$. Enter the results below, along with the corresponding results found in Table A-1 of the textbook.

x	$P(x)$ from STATDISK	$P(x)$ from Table A-1
0		
1		
2		
3		
4		

By comparing the above results, what advantage does STATDISK have over Table A-1?

____

5-5. ***Binomial Probabilities*** Consider binomial probabilities corresponding to $n = 4$ and $p = 1/4$ (or 0.25). If you attempt to use Table A-1 for finding the probabilities, you will find that the table does not apply to a probability of $p = 1/4$. Use STATDISK to find the probabilities and enter the results below.

x	$P(x)$
0	
1	
2	
3	
4	

5-6. ***Binomial Probabilities*** Assume that boys and girls are equally likely and 100 births are randomly selected. Use STATDISK with $n = 100$ and $p = 0.5$ to find $P(x)$, where x represents the number of girls among the 100 babies.

a. $x = 35$ ________

b. $x = 45$ ________

c. $x = 50$ ________

5-7. ***Binomial Probabilities*** We often assume that boys and girls are equally likely, but the actual values are P(boy) = 0.5121 and P(girl) = 0.4879. Repeat Experiment 5-6 using these values, then compare these results to those obtained in Experiment 5-6.

a. $x = 35$ _____ ____

b. $x = 45$ _____ ____

c. $x = 50$ _____ ____

5-8. ***Cumulative Probabilities*** Assume that P(boy) = 0.5121, P(girl) = 0.4879, and that 100 births are randomly selected. Use STATDISK to find the probability that the number of girls among 100 babies is . . .

a. Fewer than 60 __________

b. Fewer than 48 __________

c. At most 30 __________

d. At least 55 __________

e. More than 40 __________

5-9. ***Identifying 0+*** In Table A-1 from the textbook, the probability corresponding to $n = 12$, $p = 0.10$, and $x = 6$ is shown as 0+. Use STATDISK to find the corresponding probability and enter the result here. ________

5-10. ***Identifying 0+*** In Table A-1 from the textbook, the probability corresponding to $n = 15$, $p = 0.80$, and $x = 5$ is shown as 0+. Use STATDISK to find the corresponding probability and enter the result here. ________

5-11. ***Identifying a Probability Distribution*** Use STATDISK to construct a table of x and $P(x)$ values corresponding to a binomial distribution in which $n = 4$ and $p = 0.3$. Enter the table in the margin.

In Experiments 5-12 through 5-20, use STATDISK for the given exercises.

5–12. ***Color Blindness*** Nine percent of men and 0.25% of women cannot distinguish between the colors red and green. This is the type of color blindness that causes problems with traffic signals. If six men are randomly selected for a study of traffic signal perceptions, find the probability that exactly two of them cannot distinguish between red and green. Enter the probability here: __________

5–13. ***Acceptance Sampling*** The Telektronic Company purchases large shipments of fluorescent bulbs and uses this acceptance sampling plan: Randomly select and test 24 bulbs, then accept the whole batch if there is only one or none that doesn't work. If a particular shipment of thousands of bulbs actually has a 4% rate of defects, what is the probability that this whole shipment will be accepted? ________

5–14. ***IRS Audits*** The Hemingway Financial Company prepares tax returns for individuals. (Motto: "We also write great fiction.") According to the Internal Revenue Service, individuals making \$25,000–\$50,000 are audited at a rate of 1%. The Hemingway Company prepares five tax returns for individuals in that tax bracket, and three of them are audited.

a. Find the probability that when 5 people making \$25,000–\$50,000 are randomly selected, exactly 3 of them are audited. ________

b. Find the probability that at least three are audited. ________

c. Based on the preceding results, what can you conclude about the Hemingway customers? Are they just unlucky, or are they being targeted for audits?

____

5–15. ***Overbooking Flights*** Air America has a policy of booking as many as 15 persons on an airplane that can seat only 14. (Past studies have revealed that only 85% of the booked passengers actually arrive for the flight.) Find the probability that if Air America books 15 persons, not enough seats will be available. Is this probability low enough so that overbooking is not a real concern for passengers? ______________________________

5–16. ***Drug Reaction*** In a clinical test of the drug Viagra, it was found that 4% of those in a placebo group experienced headaches.

a. Assuming that the same 4% rate applies to those taking Viagra, find the probability that among 8 Viagra users, 3 experience headaches. ________

b. Assuming that the same 4% rate applies to those taking Viagra, find the probability that among 8 randomly selected users of Viagra, all 8 experienced a headache. ________

c. If all 8 Viagra users were to experience a headache, would it appear that the headache rate for Viagra users is different than the 4% rate for those in the placebo group? Explain. ______________________________

5–17. ***TV Viewer Surveys*** The CBS television show *60 Minutes* has been successful for many years. That show recently had a share of 20, meaning that among the TV sets in use, 20% were tuned to *60 Minutes* (based on data from Nielsen Media Research). Assume that an advertiser wants to verify that 20% share value by conducting its own survey, and a pilot survey begins with 10 households having TV sets in use at the time of a *60 Minutes* broadcast.

a. Find the probability that none of the households are tuned to *60 Minutes.*______

b. Find the probability that at least one household is tuned to *60 Minutes.*________

c. Find the probability that at most one household is tuned to *60 Minutes.*________

d. If at most one household is tuned to *60 Minutes,* does it appear that the 20% share value is wrong? Why or why not?

____

5–18. ***Affirmative Action Programs*** A study was conducted to determine whether there were significant differences between medical students admitted through special programs (such as affirmative action) and medical students admitted through the regular admissions criteria. It was found that the graduation rate was 94% for the medical students admitted through special programs (based on data from the *Journal of the American Medical Association*).

a. If 10 of the students from the special programs are randomly selected, find the probability that at least 9 of them graduated. _________

b. Would it be unusual to randomly select 10 students from the special programs and get only 7 that graduate? Why or why not?

____

5–19. ***Identifying Gender Discrimination*** After being rejected for employment, Kim Kelly learns that the Bellevue Advertising Company has hired only two women among the last 20 new employees. She also learns that the pool of applicants is very large, with an approximately equal number of qualified men and women. Help her address the charge of gender discrimination by finding the probability of getting two or fewer women when 20 people are hired, assuming that there is no discrimination based on gender. Does the resulting probability really support such a charge?

____

5–20. ***Testing Effectiveness of Gender Selection Technique*** When testing a method of gender selection, 12 babies are born. Construct a table for the probability distribution that results from 12 births, then determine whether a gender selection technique appears to be effective if there are 9 girls and 3 boys.

____

____

____

In Experiments 5-21 through 5-24, use STATDISK to solve the given exercises involving the Poisson Distribution.

5–21. ***Radioactive Decay*** Radioactive atoms are unstable because they have too much energy. When they release their extra energy, they are said to decay. When studying Cesium 137, it is found that during the course of decay over 365 days, 1,000,000 radioactive atoms are reduced to 977,287 radioactive atoms.

a. Find the mean number of radioactive atoms lost through decay in a day. ______

b. Find the probability that on a given day, 50 radioactive atoms decayed. ______

5-22. ***Aircraft Hijackings*** For the past few years, there has been a yearly average of 29 aircraft hijackings worldwide (based on data from the FAA). The mean number of hijackings per day is estimated as $\mu = 29/365$. If the United Nations is organizing a single international hijacking response team, there is a need to know about the chances of multiple hijackings in one day. Find the probability that the number of hijackings (x) in one day is 0 or 1.

____

What do you conclude about the United Nation's organizing of a single response team?

____

5-23. ***Deaths From Horse Kicks*** A classic example of the Poisson distribution involves the number of deaths caused by horse kicks of men in the Prussian Army between 1875 and 1894. Data for 14 corps were combined for the 20-year period, and the 280 corps-years included a total of 196 deaths. After finding the mean number of deaths per corps-year, find the probability that a randomly selected corps-year has the following numbers of deaths.

a. 0 _____ b. 1 _____ c. 2 _____ d. 3 _____ e. 4 _____

The actual results consisted of these frequencies: 0 deaths (in 144 corps-years); 1 death (in 91 corps-years); 2 deaths (in 32 corps-years); 3 deaths (in 11 corps-years); 4 deaths (in 2 corps-years). Compare the actual results to those expected from the Poisson probabilities. Does the Poisson distribution serve as a good device for predicting the actual results?

____

5-24. ***Homicide Deaths*** In one year, there were 116 homicide deaths in Richmond, Virginia (based on "A Classroom Note On the Poisson Distribution: A Model for Homicidal Deaths In Richmond, VA for 1991," *Mathematics and Computer Education,* by Winston A. Richards). For a randomly selected day, find the probability that the number of homicide deaths is

a. 0 _____ b. 1 _____ c. 2 _____ d. 3 _____ e. 4 _____

Compare the calculated probabilities to these actual results: 268 days (no homicides); 79 days (1 homicide); 17 days (2 homicides); 1 day (3 homicides); there were no days with more than 3 homicides.____

____

6

Normal Distributions

6-1 Randomly Generating Normally Distributed Data

6-2 Probabilities and Values from a Normal Distribution

6-3 The Central Limit Theorem

6-4 Assessing Normality

6-1 Randomly Generating Normally Distributed Data

There are many cases in which real data have a distribution that is approximately normal, and we can learn much about normal distributions by studying such cases. But how do we know when an "approximately" normal distribution is too far away from a perfect normal distribution? One way to circumvent this issue is to sample from a population with a known normal distribution and known parameters. Although this is often difficult to accomplish in reality, we can use the wonderful power of computers to obtain samples from theoretical normal distributions, and STATDISK has such a capability.

Consider IQ scores. IQ tests are designed to produce a mean of 100 and a standard deviation of 15, and we expect that such scores are normally distributed. Suppose we want to learn about the variation of sample means for samples of IQ scores. Instead of randomly selecting groups of people and giving them IQ tests, we can sample from theoretical populations. We can then learn much about the distribution of sample means. The following procedure allows you to obtain a random sample from a normally distributed population with a given mean and standard deviation.

STATDISK Procedure for Randomly Generating Sample Values from a Normally Distributed Population

1. Select **Data** from the main menu at the top of the screen.

2. Select **Normal Generator** from the subdirectory.

3. You will now get a dialog box, so enter the following.

   -Enter the desired sample size (such as 500) for the number of values to be generated.
   -Enter the desired mean (such as 100).
   -Enter the desired standard deviation (such as 15).
   -Enter the desired number of decimal places for the generated values.
   -Enter a number for the Random Seed only if you want to *repeat* the generation of the data set. Otherwise, leave that box empty. (Leaving the Random Seed box empty causes a different data set to be randomly generated each time; if you use the same number as the seed, you will generate the same data set each time.)

4. Click on the **Generate** button.

Shown on the next page is the dialog box for generating 500 sample values (with 0 decimal places) from a normally distributed population with a mean of 100 and a standard deviation of 15. After clicking on **Generate**, the generated values will appear in the column at the right. You can then click on **Copy** so that the 500 sample values can be copied to the Statdisk data window, where the values can be used for programs such as those for creating a histogram, boxplot, or calculating the descriptive statistics.

The result of this process is a collection of *sample* data randomly generated from a population with the specified mean and standard deviation, so the mean of the sample data might not be exactly the same as the value specified, and the standard deviation of the sample data might not be exactly the same as the value specified. The sample of IQ scores generated in this case has a mean of 100.73 and a standard deviation of 15.416.

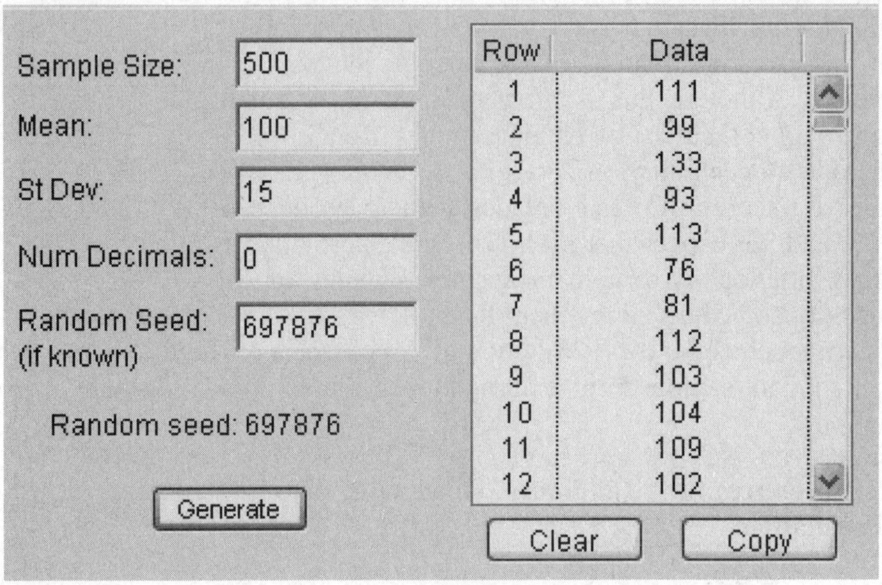

Sample Size: 500
Mean: 100
St Dev: 15
Num Decimals: 0
Random Seed: (if known) 697876
Random seed: 697876
Generate

Row	Data
1	111
2	99
3	133
4	93
5	113
6	76
7	81
8	112
9	103
10	104
11	109
12	102

Clear Copy

6-2 Probabilities and Values from a Normal Distribution

The "Normal Probability Distributions" chapter in the textbook (Chapter 6 in *Elementary Statistics*, 10th edition) begin with sections discussing the standard normal distribution and applications of normal distributions. (Recall that a standard normal distribution has a mean of 0 and a standard deviation of 1.) Table A-2 in Appendix B of the textbook lists a variety of different z scores along with their corresponding areas. STATDISK's Normal Probability function can be used in place of Table A-2. STATDISK is much more flexible than the table, and it includes many more values.

The textbook notes that when working with a normal distribution with a mean that is not 0 or a standard deviation that is not 1, the equation $z = (x - \mu)/\sigma$ can be used to convert between x values and z scores. Here is the procedure for using STATDISK:

STATDISK Procedure for Finding Probabilities or z Scores with a Normal Distribution

1. Select **Analysis** from the main menu at the top of the screen.
2. Select **Probability Distributions** from the subdirectory.
3. Select **Normal Distribution**.
4. Either enter the z score or enter the cumulative area from the left, then click **Evaluate**.

If you enter a z score in Step 4, the display will include corresponding *areas*. If you enter the cumulative area from the left, the display will include the corresponding z score (along with other areas). For example, using the above procedure, enter a z score of 1 and click Evaluate. The screen display will be as shown below.

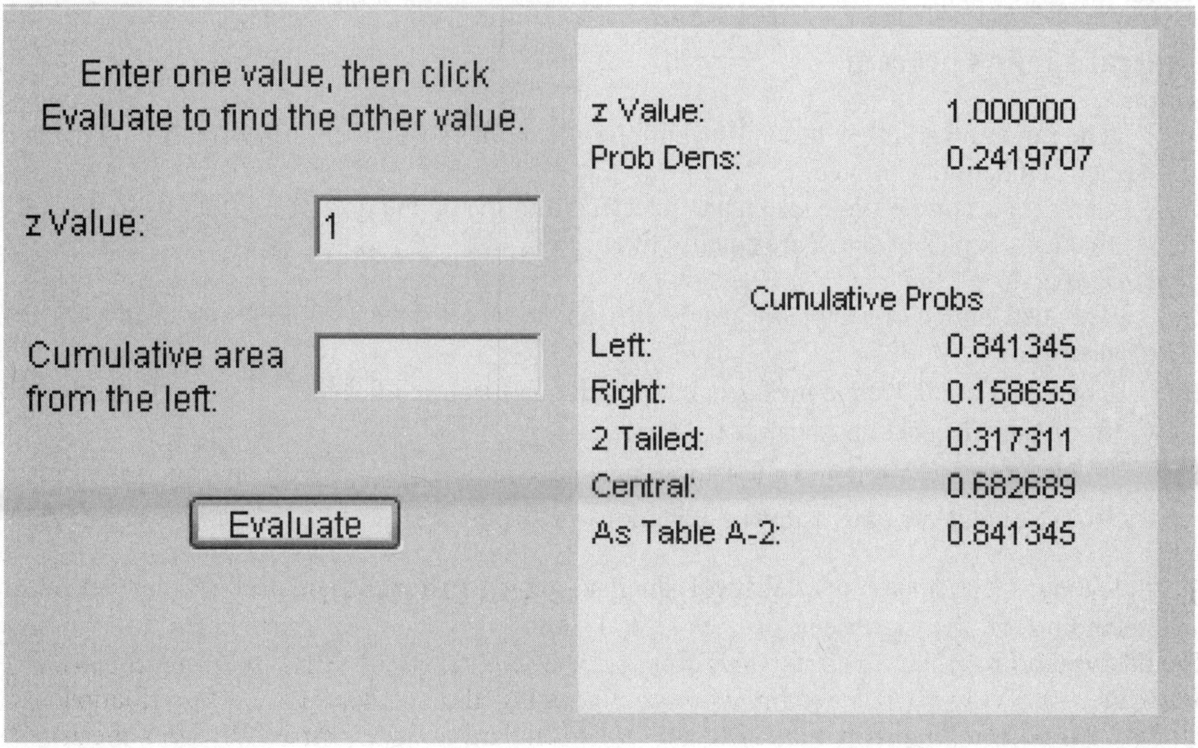

Enter one value, then click Evaluate to find the other value.

z Value: 1

Cumulative area from the left:

Evaluate

z Value:	1.000000
Prob Dens:	0.2419707

Cumulative Probs

Left:	0.841345
Right:	0.158655
2 Tailed:	0.317311
Central:	0.682689
As Table A-2:	0.841345

The following areas are included in the above STATDISK display, and they correspond to the entry of $z = 1$.

Left	Area below the normal curve and to the left of $z = 1$:	0.841345
Right	Area below the normal curve and to the right of $z = 1$:	0.158655
2 Tailed	Twice the area in the tail bounded by $z = 1$:	0.317311
Central	Twice the area below the curve and bounded by the centerline and $z = 1$:	0.682689
As Table A-2	The area below the curve and to the *left* of $z = 1$: (The label "As Table A-2" indicates that the values are based on *cumulative areas from the left*, as in Table A-2.)	0.841345

[The value shown for "Prob Dens" (probability density) is the height of the normal distribution curve for the value of z. The above display shows that when $z = 1$, the graph of the standard normal distribution has a height of 0.2419707. This particular value is not used in the textbook.]

6-3 The Central Limit Theorem

The textbook discusses the central limit theorem, which is applied often throughout the remaining chapters of the textbook. From the textbook we have the following statement of the central limit theorem.

Central Limit Theorem

Given:

1. The random variable x has a distribution (which may or may not be normal) with mean μ and standard deviation σ.

2. Samples all of the same size n are selected from the population of x values in such a way that all samples of size n are equally likely.

Conclusions:

1. The distribution of sample means $\bar{x}$ will, as the sample size increases, approach a *normal* distribution.

2. The mean of all sample means is the population mean μ. (That is, the normal distribution from Conclusion 1 has mean μ.)

3. The standard deviation of all sample means is $\sigma / \sqrt{n}$. (That is, the normal distribution from Conclusion 1 has standard deviation $\sigma / \sqrt{n}$.)

In the study of methods of statistical analysis, it is extremely helpful to have a clear understanding of the statement of the central limit theorem. See Experiment 6–3 in this manual/workbook.

6–4 Assessing Normality

The textbook uses the following criteria for determining whether sample data appear to come from a population having a normal distribution. (The section "Assessing Normality is not included in *Essentials of Statistics*.)

1. **Histogram:** Construct a Histogram. Reject normality if the histogram departs dramatically from a bell shape. STATDISK can generate a histogram.

2. **Outliers:** Identify outliers. Reject normality if there is more than one outlier present. (Just one outlier could be an error or the result of chance variation, but be careful, because even a single outlier can have a dramatic effect on results.) Using STATDISK, we can use the "Data tools" button to sort the data and easily identify any values that are far away from the majority of all other values.

3. **Normal Quantile Plot:** If the histogram is basically symmetric and there is at most one outlier, construct a *normal quantile plot.* Examine the normal quantile plot and reject normality if the points do not lie close to a straight line, or if the points exhibit some systematic pattern that is not a straight–line pattern. STATDISK can generate a normal quantile plot. (Select **Data**, then **Normal Quantile Plot**.)

As an example, consider the 52 rainfall amounts in Boston for the Sundays in one year. (The Sunday rainfall amounts in Boston are included as a STATDISK data set.) Shown below are the histogram and normal quantile plot. The histogram is far from being bell–shaped, which suggests that the Sunday rainfall amounts do not have a normal distribution. The points in the normal quantile plot do not lie close to a straight line, further suggesting that the Sunday rainfall amounts are not normally distributed. Also, we can investigate outliers by sorting the data. (Examining the sorted list of 52 rainfall amounts, we find that 1.28 in. appears to be the only outlier. Because there is only one outlier, we make no conclusion about the normality of the data based on outliers.) After considering the shape of the histogram, the presence of outliers, and the normal quantile plot, we conclude that the sample rainfall amounts appear to come from a population having a distribution that is *not* a normal distribution.

STATDISK–Generated Histogram

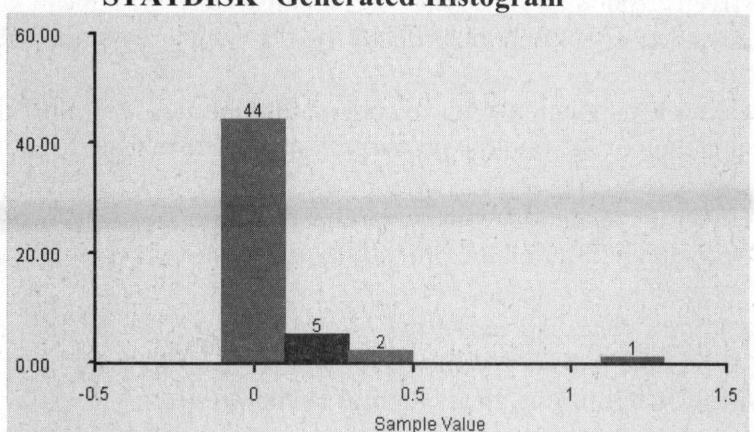

STATDISK–Generated Normal Quantile Plot

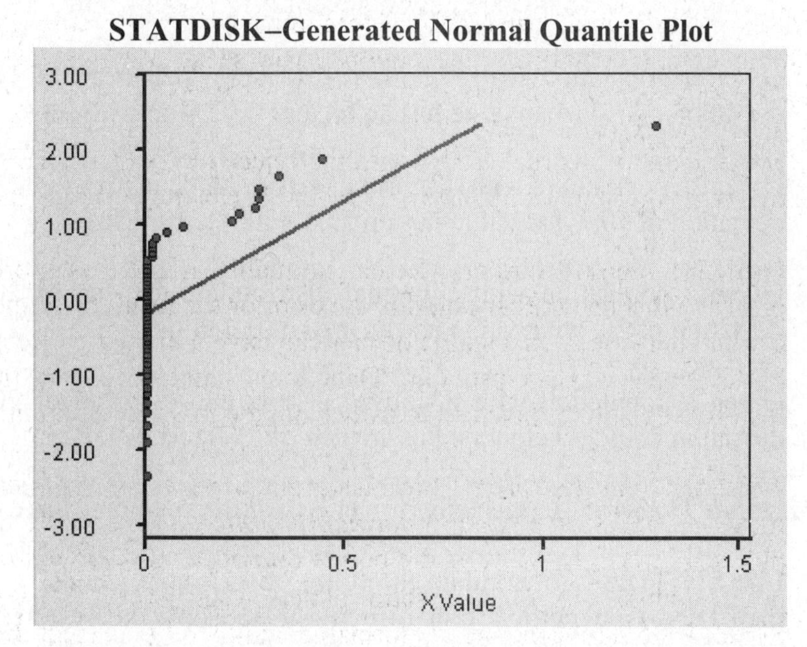

CHAPTER 6 EXPERIMENTS: Normal Distributions

6-1. ***Finding Probabilities for a Normal Distribution*** Use STATDISK's **Normal Distribution module** to find the indicated probabilities. First select **Analysis** from the main menu, then select **Probability Distributions**, then **Normal Distribution**.

a. Given a population with a normal distribution, a mean of 0, and a standard deviation of 1, find the probability of a value less than 1.50.________________

b. Given a population with a normal distribution, a mean of 100, and a standard deviation of 15, find the probability of a value less than 120.________________

c. Given a population with a normal distribution, a mean of 75, and a standard deviation of 10, find the probability of a value greater than 80.______________

d. Given a population with a normal distribution, a mean of 200, and a standard deviation of 20, find the probability of a value between 180 and 210.__________

e. Given a population with a normal distribution, a mean of 200, and a standard deviation of 20, find the probability of a value between 205 and 223.__________

6-2. ***Finding Values for a Normal Distribution*** Use STATDISK's **Normal Distribution** module to find the indicated values. First select **Analysis** from the main menu, then select **Probability Distributions**, then **Normal Distribution**.

a. Given a population with a normal distribution, a mean of 0, and a standard deviation of 1, what value has an area of 0.1234 to its left?________________

b. Given a population with a normal distribution, a mean of 100, and a standard deviation of 15, what value has an area of 0.1234 to its right?_______________

c. Given a population with a normal distribution, a mean of 75, and a standard deviation of 10, what value has an area of 0.9 to its left?__________________

d. Given a population with a normal distribution, a mean of 200, and a standard deviation of 20, what value has an area of 0.9 to its right?________________

e. Given a population with a normal distribution, a mean of 200, and a standard deviation of 20, what value has an area of 0.005 to its left?________________

6-3. ***Central Limit Theorem*** In this experiment, assume that all dice have 6 sides.

a. Use STATDISK to simulate the rolling of a single die 800 times. (Select **Data**, then **Dice Generator**.) Use Copy/Paste to copy the results to the Descriptive Statistics and Histogram modules, and enter the actual results below.

One Die: Mean: ________
Standard Deviation: ________
Distribution shape: ________

b. Part (a) used a single die, but we will now use a pair of dice. Use STATDISK to "roll" two dice 800 times (again using Data/Dice Generator). The 800 values are *totals* for each pair of dice, so transform the totals to *means* by dividing each total by 2. (Use Copy/Paste to copy the results to the Statdisk data window, then use STATDISK's **Sample Transformations** feature to divide each value by 2. To divide each value by 2, select the operation of / and use a constant of 2.) Now use Copy/Paste to copy the 800 *means* to the Statdisk data window, then use the Descriptive Statistics and Histogram modules. Enter the results below.

Two Dice: Mean: ________
Standard Deviation: ________
Distribution shape: ________

c. Repeat part (b) using 10 dice. When finding the mean of the 10 dice, divide each value by 10.

10 Dice: Mean: ________
Standard Deviation: ________
Distribution shape: ________

d. Repeat part (b) using 20 dice.

20 Dice: Mean: ________
Standard Deviation: ________
Distribution shape: ________

e. General conclusions:
What happens to the mean as the sample size increases from 1 to 2 to 10 to 20?

____

What happens to the standard deviation as the sample size increases?

____

What happens to the distribution shape as the sample size increases?

____

How do these results illustrate the central limit theorem?

____

____

6-4. ***Determining Significance*** People generally believe that the mean body temperature is 98.6°F. The Body Temperatures data set in Appendix B of the textbook includes a sample of 106 body temperatures with these properties: The distribution is approximately normal, the sample mean is 98.20°F, and the standard deviation is 0.62°F. We want to determine whether these sample results differ from 98.6°F by a *significant* amount. One way to make that determination is to study the behavior of samples drawn from a population with a mean of 98.6.

a. Use STATDISK to generate 106 values from a normally distributed population with a mean of 98.6 and a standard deviation of 0.62. Use STATDISK to find the mean of the generated sample. Record that mean here:____

b. Repeat part (a) nine more times and record the 10 sample means here:

____

c. By examining the 10 sample means in part (b), we can get a sense for how much sample means vary for a normally distributed population with a mean of 98.6 and a standard deviation of 0.62. After examining those 10 sample means, what do you conclude about the likelihood of getting a sample mean of 98.20? Is 98.20 a sample mean that could easily occur by chance, or is it significantly different from the likely sample means that we expect from a population with a mean of 98.6?

____

d. Given that researchers did obtain a sample of 106 temperatures with a mean of 98.20°F, what do their results suggest about the common belief that the population mean is 98.6°F?

____

6-5. ***Determining Significance*** The U.S. Department of the Treasury claims that the procedure it uses to mint quarters yields a mean weight of 5.670 g.

a. Refer to the Coin Weights data set in Appendix B of the textbook and find the weights of the 40 quarters produced after 1964. Find the mean and standard deviation of those weights and enter them below.

Sample mean:____________ Standard deviation:____________

b. Generate 10 different samples, where each sample has 40 values randomly selected from a normally distributed population with a mean of 5.670 g and a standard deviation of 0.068 g (based on the U.S. Department of the Treasury specifications). For each sample, record the sample mean and enter it here.

____

(*continued*)

c. By examining the 10 sample means in part (b), we can get a sense for how much sample means vary for a normally distributed population with a mean of 5.670 and a standard deviation of 0.068. After examining those 10 sample means, what do you conclude about the likelihood of getting a sample mean like the one found for the sample of quarters? Is the sample mean a value that could easily occur by chance, or is it significantly different from the likely sample means that we expect from a population with a mean of 5.670?

____

d. Consider the sample mean found from the data set in Appendix B from the textbook. Does it suggest that the population mean of 5.670 g is not correct?

____

Assessing Normality *In Experiments 6-6 through 6-8, refer to the indicated STATDISK data set. In each case, print a histogram, print a normal quantile plot, and identify any outliers. Based on the results, determine whether the sample data appear to come from a normally distributed population.*

6-6. **Boston Rainfall**: Use the amounts of rainfall in Boston on Fridays.

Outliers:________________________

Normal Distribution? (Give reasons.) ______________________________

____

6-7. **Cans**: Use the axial loads of aluminum cans that are 0.0111 in. thick.

Outliers:________________________

Normal Distribution? (Give reasons.) ______________________________

____

6-8. **Homeruns**: Use the distances of homeruns hit by Mark McGwire in 1998.

Outliers:________________________

Normal Distribution? (Give reasons.) ______________________________

____

Confidence Intervals and Sample Sizes

7-1 Confidence Intervals for Estimating *p*

The textbook introduces confidence intervals as a tool for estimating a population proportion p. (See Section 7-2 in *Elementary Statistics*, 10th edition.) The following definition was included.

Definition
A **confidence interval** (or **interval estimate**) is a range (or an interval) of values used to estimate the true value of a population parameter.

When finding a confidence interval estimate of p, STATDISK requires the sample size n and the number of successes x. In some cases, the values of x and n are both known, but in other cases the given information may consist of n and a sample percentage. For example, consider the statement that "a survey shows that among 1050 randomly selected voters, 59.7% favor the Republican candidate for the presidency." Based on that information, we know that $n = 1050$ and $\hat{p} = 0.597$, but we don't know the value of x. Because $\hat{p} = x/n$, it follows that $x = \hat{p}n$, so the number of successes can be found by multiplying the sample proportion p and the sample size n. With $n = 1050$ and $\hat{p} = 0.597$, we get $x = (0.597)(1050) = 626.85$. Because x must be a whole number, we let $x = 627$.

To find the number of successes x from the sample proportion and sample size: Calculate $x = \hat{p}n$, and round the result to the nearest whole number.

After identifying the sample size n and the number of successes x, proceed as follows.

STATDISK Procedure for Finding Confidence Intervals for *p*

1. Select **Analysis** from the main menu.

2. Select **Confidence Intervals**.

3. Select **Population Proportion**.

4. Make these entries in the dialog box:

   -Enter a confidence level, such as 0.95 or 0.99.
   -Enter the value for the sample size n.
   -Enter the number of successes for x.

5. Click on the **Evaluate** button.

Shown below is the STATDISK display resulting from entries of 0.95 for a confidence level, 1050 for the sample size n, and 627 for the number of successes x. The 95% confidence interval is $0.567 < p < 0.627$ (rounded). This can also be expressed as $56.7\% < p < 62.7\%$ or as $59.7\% \pm 3.0\%$. The media would typically report this result as "60% with a margin of error of three percentage points."

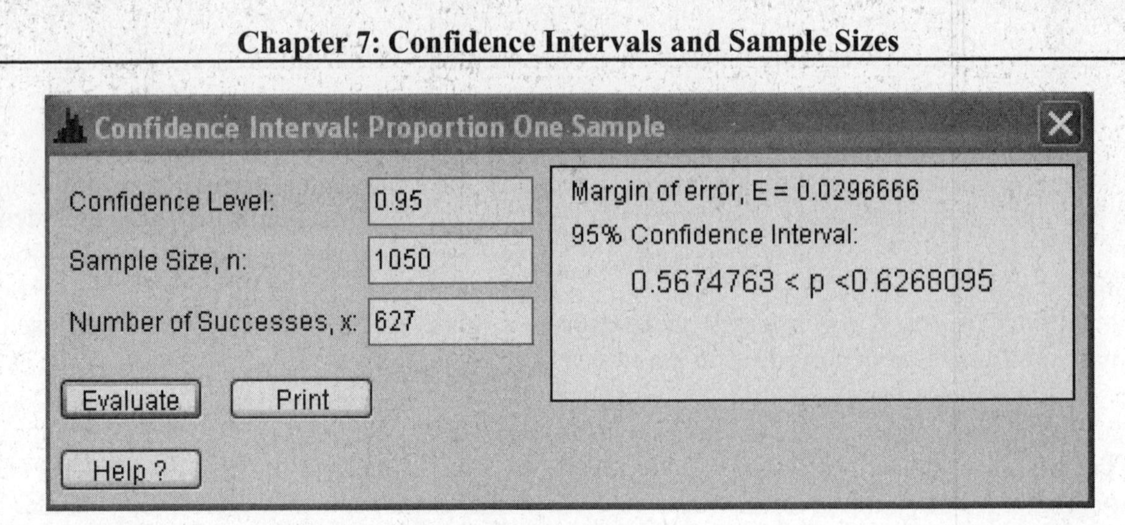

7-2 Confidence Intervals for Estimating μ

The textbook discuss the construction of confidence interval estimates of a population mean μ. (See Sections 7-3 and 7-4 in *Elementary Statistics,* 10th edition.) The textbook stresses the importance of selecting the correct distribution (normal or *t*), but STATDISK automatically chooses the correct distribution based on the information that is entered. STATDISK is very easy to use for constructing confidence interval estimates for a population mean μ. However, STATDISK requires that you first obtain the descriptive statistics of n, $\bar{x}$, and s, as indicated in Step 1 of the following procedure.

STATDISK Procedure for Finding Confidence Intervals for μ

1. If the sample data are known but n, $\bar{x}$, and s are not yet known, find the values of those sample statistics by using STATDISK's Explore Data or Descriptive Statistics features. (See Section 3-1 of this manual/workbook.) You must know the values of n, $\bar{x}$, and s before proceeding to step 2.

2. Select **Analysis** from the main menu at the top of the screen.

3. Select **Confidence Intervals** from the subdirectory.

4. Select **Population Mean.**

5. You will now see a dialog box allowing you to make these entries.
   - -Enter a Confidence Level, such as 0.95 or 0.99.
   - -Enter the Sample Size n.
   - -Enter the value of the sample mean $\bar{x}$.
   - -Enter the value of the sample standard deviation s.
   - -Enter the value of the *population* standard deviation σ if it is known. (The value of σ is usually unknown, so you will usually leave this entry blank.)

6. Click the **Evaluate** button.

To illustrate this procedure, consider this example:

> **Pulse Rates of Females** For the sample of pulse rates of women in Data Set 1 in Appendix B, we have $n = 40$, $\bar{x} = 76.3$, and $s = 12.5$, and the sample is a simple random sample. Use STATDISK with the sample data to construct a 95% confidence interval estimate of the population mean μ.

Using STATDISK with $n = 40$, $\bar{x} = 76.3$, and $s = 12.5$, we obtain the following display. This display shows the confidence interval of $72.3 < \mu < 80.3$ (rounded). With a margin of error of 4.0 (rounded), we can also express the confidence interval as 76.3 ± 4.0.

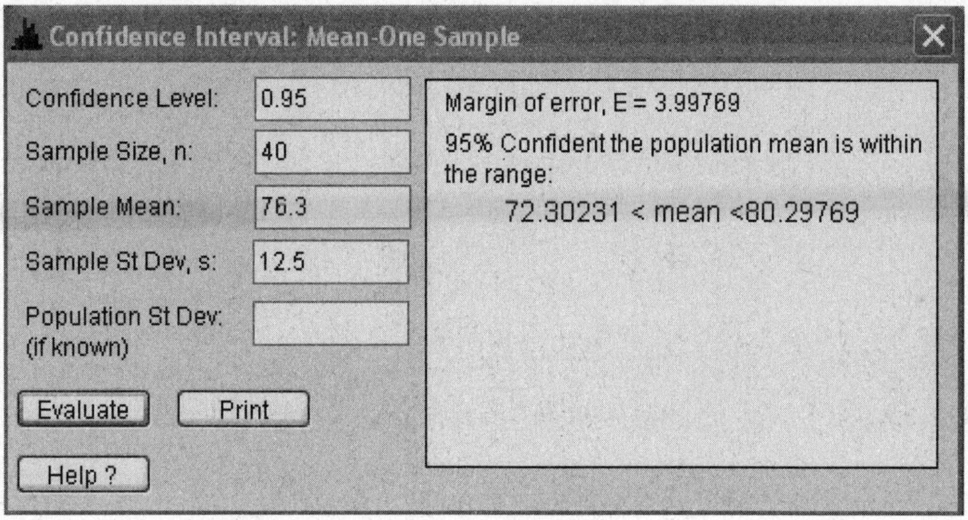

Confidence Interval: Mean-One Sample

Confidence Level: 0.95
Sample Size, n: 40
Sample Mean: 76.3
Sample St Dev, s: 12.5
Population St Dev: (if known)
Evaluate Print
Help ?

Margin of error, E = 3.99769
95% Confident the population mean is within the range:
72.30231 < mean <80.29769

7-3 Confidence Intervals for Estimating σ

The construction of confidence interval estimates of a population standard deviation σ or variance σ^2 is described in the textbook. (See Section 7-5 in *Elementary Statistics*, 10th edition.) After selecting a confidence level and entering the sample size n and sample standard deviation s, STATDISK will automatically provide a confidence interval estimate of σ along with a confidence interval estimate of σ^2. You get both confidence intervals (for σ and σ^2), whether you want them or not. Be careful to correctly identify the value of the sample standard deviation s. Be careful to enter the sample standard deviation where it is required; if only the sample variance is known, find its square root and enter that value for s. After obtaining the values of the sample size n and sample standard deviation s, proceed as follows.

STATDISK Procedure for Finding Confidence Intervals for σ and σ^2

1. Select **Analysis** from the main menu.

2. Select **Confidence Intervals**.

3. Select **Population St Dev**.

4. Make these entries in the dialog box:
   -Enter a confidence level, such as 0.95 or 0.99.
   -Enter the value for the sample size *n.*
   -Enter the value of the sample *standard deviation s* (not the *variance* s^2)

5. Click on the **Evaluate** button.

7-4 Sample Sizes for Estimating *p*

The textbook describes methods for determining the *sample size* needed to estimate a population proportion *p*. (See Sectioin 7-2 in *Elementary Statistics*, 10th edition.) STATDISK requires that you enter a confidence level (such as 0.95) and a margin of error *E* (such as 0.03). In addition to those two required entries, there are also two optional entries. You can enter an estimate of *p* if one is known, based on such factors as prior knowledge or results from a previous study. You can also enter the population size *N* if it is known and if you are sampling without replacement. The textbook includes the following two cases.

When an Estimate $\hat{p}$ Is Known: Formula 7-2 $n = \dfrac{\left[z_{\alpha/2}\right]^2 \hat{p}\hat{q}}{E^2}$

When No Estimate $\hat{p}$ Is Known: Formula 7-3 $n = \dfrac{\left[z_{\alpha/2}\right]^2 \cdot 0.25}{E^2}$

STATDISK Procedure for Finding Sample Sizes Required to Estimate *p*

1. Select **Analysis** from the main menu.

2. Select the subdirectory item of **Sample Size Determination**.

3. Select **Estimate Proportion**.

4. Make these entries in the dialog box:
   -Enter a confidence level, such as 0.95 or 0.99.
   -Enter a margin of error *E*. (*Hint:* The margin of error must be expressed in decimal form. For example, a margin of error of "three percentage points" should be entered as 0.03.)
   -Enter an estimated proportion if it is known. (This value might come from a previous study, or from knowledge about the value of the sample proportion. If such a value is not known, leave this box empty.)
   -Enter a value for the population size *N* if you will sample without replacement from a finite population of *N* subjects. (If the population is large or sampling is done with replacement, leave this box blank.)

5. Click the **Evaluate** button.

7-5 Sample Sizes for Estimating μ

The textbook discusses the procedure for determining the sample size necessary to estimate a population mean μ. (See Section 7-3 in *Elementary Statistics,* 10th edition.) We use the formula

$$n = \left[\frac{z_{\alpha/2}\sigma}{E}\right]^2$$

STATDISK requires that we know the desired degree of confidence, the margin of error E, and the population standard deviation σ. The textbook notes that it is unusual to know σ without knowing μ, but σ might be known from a previous study or it might be estimated from a pilot study or the range rule of thumb. The entry of a finite population size N is optional, as described in the following steps.

STATDISK Procedure for Finding Sample Sizes Required to Estimate μ

1. Select **Analysis** from the main menu.

2. Select **Sample Size Determination** from the subdirectory.

3. Select the option of **Estimate Mean**.

4. In the dialog box, make these entries:
   - -Enter a confidence level, such as 0.95 or 0.99.
   - -Enter the value of the population standard deviation σ. (If σ is not known, consider estimating it from a previous study or pilot study or use the range rule of thumb.)
   - -For the entry box labeled Population Size N: Leave it blank if you are sampling with replacement, or if you have a small sample drawn from a large population. (Consider a sample size n to be "small" if $n \leq 0.05N$.) Enter a value only if you are sampling without replacement from a finite population with known size N, and the sample is large so that $n > 0.05N$. This box is usually left blank.

5. Click on the **Evaluate** button.

7-6 Sample Sizes for Estimating σ

The textbook describes a procedure for determining the sample size required to estimate a population standard deviation σ or population variance σ^2. (See Section 7-5 in *Elementary Statistics*, 10th edition.) Table 7-2 in the textbook lists sample sizes for several different cases, but STATDISK is much more flexible and allows you to find sample sizes for many other cases.

STATDISK Procedure for Finding Samples Sizes Required to Estimate σ or σ^2

1. Select **Analysis** from the main menu.

2. Select **Sample Size Determination**.

3. Select the third option of **Estimate St Dev**.

4. Make these entries in the dialog box:
   -Enter a confidence level, such as 0.95 or 0.99.
   -Enter a "Percent Margin of Error, %E." (For example, if you enter 20, STATDISK will provide the sample size required so that s is within 20% of σ, and it will also provide the sample size required so that s^2 is within 20% of σ^2.)

5. Click the **Evaluate** button.

Statistics textbooks tend to omit discussions about the issue of determining sample size for estimating a population standard deviation σ or variance σ^2, but STATDISK now allows you to do calculations with ease.

7-7 Bootstrap Resampling

Suppose that we have sample data consisting of the values 27, 31, 32, 35, and 200, and we want to use those values to construct a confidence interval estimate of the population mean. That small sample of five values includes the value of 200, which is an outlier. Shown below is the normal quantile plot of these values. Because the points do not fit a straight-line pattern reasonably well, we conclude that the data values are from a distribution that is not normal.

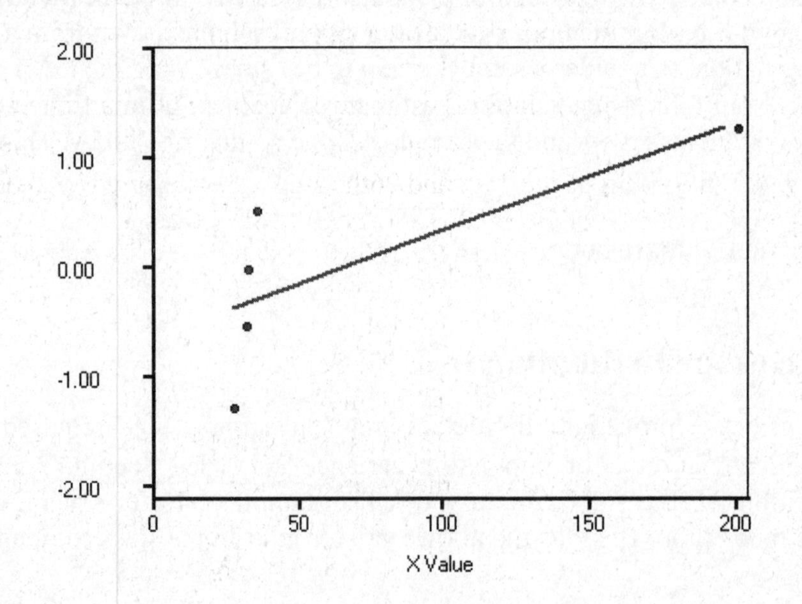

The sample is small, and σ is unknown, and the distribution is not normal, so the methods of Chapter 7 in the textbook cannot be used. One alternative is to use the method of **bootstrap resampling**, which does not require normally distributed data. With bootstrap resampling of 5 sample values, we randomly select 5 of those values, but we sample with replacement. Then we repeat that selection process many times, thereby pulling the original sample up "by its own bootstraps."

Here is a bootstrap resampling procedure for using a small sample (from a non-normal population) to find an estimate of the population mean:

1. Generate a large number of new samples by randomly selecting values (with replacement) from the original sample.
2. Find the means of the generated samples.
3. Sort the sample means.
4. Construct a 95% confidence interval estimate of the population mean μ by finding the percentiles $P_{2.5}$ and $P_{97.5}$ for the list of sorted means. Those percentile values are the confidence interval limits.

STATDISK Procedure for Bootstrap Resampling

1. Enter the sample data in column 1 of the Statdisk data window.

2. Select **Analysis**, then select the menu item of **Bootstrap Resampling**.

3. Enter the desired number of samples to be generated, such as 1000, then click **Resample**.

4. The sample means will be inserted in column 2 of the Statdisk data window, and the sample standard deviations will be inserted in column 3. Click on the **Data tools** button and proceed to sort column 2. Also sort column 3. (*Hint:* Sort columns 2 and 3 separately by using the option of "Sort a single column.")

5. Construct a 95% confidence interval estimate of the population mean μ by finding the percentiles $P_{2.5}$ and $P_{97.5}$ for the list of sorted means. (With 1000 sorted sample means, $P_{2.5}$ is the mean of the 25th and 26th sample means, and $P_{97.5}$ is the mean of the 975th and 976th sample means.) Those percentile values are the confidence interval limits. A confidence interval estimate of σ can also be found by using the same basic procedure.

Using this procedure with the sample values of 27, 31, 32, 35, and 200, the 95% confidence interval of $29.4 < \mu < 132.6$ was obtained. If the methods of Section 7-4 in the textbook are used, the confidence interval of $-28.7 < \mu < 158.8$ is obtained. Because the methods of Section 7-4 should not be used (because the requirements are not met), this latter confidence interval should not be used.

CHAPTER 7 EXPERIMENTS: Confidence Intervals and Sample Sizes

In Experiments 7–1 through 7–4, use STATDISK with the sample data and confidence level to construct the confidence interval estimate of the population proportion p.

7–1. $n = 500$, $x = 200$, 95% confidence ____________________

7–2. $n = 1200$, $x = 800$, 99% confidence ____________________

7–3. $n = 1068$, $x = 267$, 98% confidence ____________________

7–4. $n = 4500$, $x = 2925$, 90% confidence ____________________

7–5. ***Internet Shopping*** In a Gallup poll, 1025 randomly selected adults were surveyed and 29% of them said that they used the Internet for shopping at least a few times a year.

a. Find the point estimate of the percentage of adults who use the Internet for shopping. __________

b. Find a 99% confidence interval estimate of the percentage of adults who use the Internet for shopping. ____________________

c. Based on the result from part (b), if a traditional retail store wants to estimate the percentage of adult Internet shoppers in order to determine the maximum impact of Internet shoppers on its sales, what percentage of Internet shoppers should be used? __________

7–6. ***Death Penalty Survey*** In a Gallup poll, 491 randomly selected adults were asked whether they are in favor of the death penalty for a person convicted of murder, and 65% of them said that they were in favor.

a. Find the point estimate of the percentage of adults who are in favor of this death penalty. __________

b. Find a 95% confidence interval estimate of the percentage of adults who are in favor of this death penalty. ____________________

c. Can we safely conclude that the majority of adults are in favor of this death penalty? Explain.

____

7–7. ***Mendelian Genetics*** When Mendel conducted his famous genetics experiments with peas, one sample of offspring consisted of 428 green peas and 152 yellow peas.

a. Use STATDISK to find the following confidence interval estimates of the percentage of yellow peas.

99% confidence interval: ____________________
98% confidence interval: ____________________
95% confidence interval: ____________________
90% confidence interval: ____________________

b. After examining the pattern of the above confidence intervals, complete the following statement. "As the degree of confidence decreases, the confidence interval limits
____."

c. In your own words, explain why the preceding completed statement makes sense. That is, why should the confidence intervals behave as you have described?

____

____

7–8. ***Misleading Survey Responses*** In a survey of 1002 people, 701 said that they voted in a recent presidential election (based on data from ICR Research Group). Voting records show that 61% of eligible voters actually did vote. Find a 99% confidence interval estimate of the proportion of people who say that they voted. ____________________

7–9. ***Estimating Car Pollution*** In a sample of seven cars, each car was tested for nitrogen-oxide emissions (in grams per mile) and the following results were obtained: 0.06, 0.11, 0.16, 0.15, 0.14, 0.08, 0.15 (based on data from the Environmental Protection Agency). Assuming that this sample is representative of the cars in use, construct a 98% confidence interval estimate of the mean amount of nitrogen-oxide emissions for all cars.

____

7–10. ***Monitoring Lead in Air*** Listed below are measured amounts of lead (in micrograms per cubic meter or $\mu g/m^3$) in the air. The Environmental Protection Agency has established an air quality standard for lead: 1.5 $\mu g/m^3$. The measurements shown below were recorded at Building 5 of the World Trade Center site on different days immediately following the destruction caused by the terrorist attacks of September 11, 2001. After the collapse of the two World Trade buildings, there was considerable concern about the quality of the air. Use the given values to construct a 95% confidence interval estimate of the mean amount of lead in the air. Is there anything about this data set suggesting that the confidence interval might not be very good? Explain.

5.40 1.10 0.42 0.73 0.48 1.10

____

7–11. ***Forecast and Actual Temperatures*** Appendix B of the textbook includes a list of actual high temperatures and the corresponding list of three-day forecast high temperatures. (The data set is named "Forecast and Actual Temps." If the difference for each day is found by subtracting the three-day forecast high temperature from the actual high temperature, the result is a list of 35 values with a mean of –1.3° and a standard deviation of 4.7°.

a. Construct a 99% confidence interval estimate of the mean difference between all actual high temperatures and three-day forecast high temperatures.____________

b. Does the confidence interval include 0°? If a meteorologist claims that three-day forecast high temperatures tend to be too high because the mean difference of the sample is –1.3°, does that claim appear to be valid? Why or why not?

____

7–12. ***Weights of Bears*** The health of the bear population in Yellowstone National Park is monitored by periodic measurements taken from anesthetized bears. A sample of 54 bears has weights listed in the Bears data set in Appendix B of the textbook.

a. Assuming that σ is known to be 121.8 lb, find a 99% confidence interval estimate of the mean of the population of all such bear weights. ____________

b. Find a 99% confidence interval estimate of the mean of the population of all such bear weights, assuming that the population standard deviation σ is not known.

____

c. Compare the results from parts (a) and (b).

____

7–13. ***Credit Rating*** When consumers apply for credit, their credit is rated using FICO (Fair, Isaac, and Company) scores. Credit ratings are given below for a sample of applicants for car loans.

661 595 548 730 791 678 672 491 492 583 762 624 769 729 734 706

Use the sample data to construct a 99% confidence interval for the mean FICO score of all applicants for credit. ____________________________

If one bank requires a credit rating of at least 620 for a car loan, does it appear that almost all applicants will have suitable credit ratings? Why or why not?

____

____

7–14. ***World's Smallest Mammal*** The world's smallest mammal is the bumblebee bat, also known as the Kitti's hog-nosed bat (or Craseonycteris thonglongyai). Such bats are roughly the size of a large bumblebee. Listed below are weights (in grams) from a sample of these bats.

1.7 1.6 1.5 2.0 2.3 1.6 1.6 1.8 1.5 1.7 2.2 1.4 1.6 1.6 1.6

Construct a 95% confidence interval estimate of their mean weight. ________________

7–15. ***Hospital Costs*** A study was conducted to estimate hospital costs for accident victims who wore seat belts. Twenty randomly selected cases have a distribution that appears to be bell-shaped with a mean of $9004 and a standard deviation of $5629 (based on data from the U.S. Department of Transportation). Construct the 99% confidence interval for the mean of all such costs. ______________

7–16. ***Cardiac Deaths*** Because cardiac deaths appear to increase after heavy snowfalls, an experiment was designed to compare cardiac demands of snow shoveling to those of using an electric snow thrower. Ten subjects cleared tracts of snow using both methods, and their maximum heart rates (beats per minute) were recorded during both activities. The following results were obtained (based on data from "Cardiac Demands of Heavy Snow Shoveling," by Franklin et al., *Journal of the American Medical Association*, Vol. 273, No. 11):

Manual Snow Shoveling Maximum Heart Rates: $n = 10$, $\bar{x} = 175$, $s = 15$

Electric Snow Thrower Maximum Heart Rates: $n = 10$, $\bar{x} = 124$, $s = 18$

a. Find the 95% confidence interval estimate of the population mean for those people who shovel snow manually.______________________

b. Find the 95% confidence interval estimate of the population mean for those people who use the electric snow thrower. ________________________

c. Compare the results from parts (a) and (b).

____

7-17. ***Confidence Interval for Estimating a Mean*** The COLA data set in Appendix B of the textbook lists sample weights of the cola in cans of regular Coke. Open the STATDISK file and use it to answer the questions that follow.

Weights (in pounds) of a sample of cans of regular Coke

0.8192	0.8150	0.8163	0.8211	0.8181	0.8247
0.8062	0.8128	0.8172	0.8110	0.8251	0.8264
0.7901	0.8244	0.8073	0.8079	0.8044	0.8170
0.8161	0.8194	0.8189	0.8194	0.8176	0.8284
0.8165	0.8143	0.8229	0.8150	0.8152	0.8244
0.8207	0.8152	0.8126	0.8295	0.8161	0.8192

a. Use STATDISK to find the following confidence interval estimates of the population mean.

99.5% confidence interval: ____________

99% confidence interval: ____________

98% confidence interval: ____________

95% confidence interval: ____________

90% confidence interval: ____________

b. Change the first weight from 0.8192 lb to 8192 lb (a common error in data entry) and find the indicated confidence intervals for the population mean.

99.5% confidence interval: ____________

99% confidence interval: ____________

98% confidence interval: ____________

95% confidence interval: ____________

90% confidence interval: ____________

c. By comparing these results from parts (a) and (b), what do you conclude about the effect of an outlier on the values of the confidence interval limits?

____

7-18. ***Simulated Data*** STATDISK is designed to generate random numbers from a variety of different sampling distributions. In this experiment we will generate 500 IQ scores, then we will construct a confidence interval based on the sample results. IQ scores have a normal distribution with a mean of 100 and a standard deviation of 15. First generate the 500 sample values as follows.

1. Click on **Data,** then select **Normal Generator**.

2. In the dialog box, enter a sample size of 500, a mean of 100, a standard deviation of 15, and enter 0 for the number of decimal places. Click **OK**.

3. Use **Data/Descriptive Statistics** to find these statistics:
$n =$ ________ $\overline{x} =$ ________ $s =$ ________

(continued)

Using the generated values, construct a 95% confidence interval estimate of the population mean of all IQ scores. Enter the 95% confidence interval here.

___

Because of the way that the sample data were generated, we *know* that the population mean is 100. Do the confidence interval limits contain the true mean IQ score of 100?

___

If this experiment were to be repeated over and over again, how often would we expect the confidence interval limits to contain the true population mean value of 100? Explain how you arrived at your answer.

___

___

7-19. ***Simulated Data*** Follow the same steps listed in Experiment 7-18 to randomly generate 500 IQ scores from a population having a normal distribution, a mean of 100, and a standard deviation of 15. Record the sample statistics here.

$n =$ ______ $\bar{x} =$ ______ $s =$ ______

Confidence intervals are typically constructed with confidence levels around 90%, 95%, or 99%. Instead of constructing such a typical confidence interval, use the generated values to construct a 50% confidence interval. Enter the result here.

___

Does the above confidence interval have limits that actually do contain the true population mean, which we know is 100?______

Repeat the above procedure 9 more times and list the resulting 50% confidence intervals here.

______ ______ ______ ______ ______

______ ______ ______ ______

Among the total of the 10 confidence intervals constructed, how many of them actually do contain the true population mean of 100? Is this result consistent with the fact that the level of confidence used is 50%? Explain.

___

___

7-20. ***M&M Weights*** Refer to the M&M data set in Appendix B and use the sample of weights of green M&M candies to construct a 95% confidence interval for the mean weight of all M&Ms. Enter the 95% confidence interval here

____

7-21. ***Estimating Standard Deviation*** Refer to the sample data used in Experiment 7-17 and use STATDISK to construct a 95% confidence interval to estimate the population standard deviation σ. Enter the result here.

____

Refer to Section 7-5 in the textbook and identify the assumptions for the procedures used to construct confidence intervals for estimating a population standard deviation.

____

____

Assuming that the sample is a simple random sample, how can the other assumption be checked? Can STATDISK be used to check for normality? If so, do such a check and report the results and conclusion here.

____

____

7–22. ***Estimating Standard Deviation*** A container of car antifreeze is supposed to hold 3785 mL of the liquid. Realizing that fluctuations are inevitable, the quality-control manager wants to be quite sure that the standard deviation is less than 30 mL. Otherwise, some containers would overflow while others would not have enough of the coolant. She selects a simple random sample, with the results given here.

3761	3861	3769	3772	3675	3861
3888	3819	3788	3800	3720	3748
3753	3821	3811	3740	3740	3839

Use these sample results to construct the 99% confidence interval for the true value of σ. Does this confidence interval suggest that the fluctuations are at an acceptable level?

____

7-23. ***Quality Control of Doughnuts*** The Hudson Valley Bakery makes doughnuts that are packaged in boxes with labels stating that there are 12 doughnuts weighing a total of 42 oz. If the variation among the doughnuts is too large, some boxes will be underweight (cheating consumers) and others will be overweight (lowering profit). A consumer would not be happy with a doughnut so small that it can be seen only with an electron microscope, nor would a consumer be happy with a doughnut so large that it resembles a tractor tire. The quality-control supervisor has found that he can stay out of trouble if the doughnuts have a mean of 3.50 oz and a standard deviation of 0.06 oz or less. Twelve doughnuts are randomly selected from the production line and weighed, with the results given here (in ounces).

3.43 3.37 3.58 3.50 3.68 3.61 3.42 3.52 3.66 3.50 3.36 3.42

Construct a 95% confidence interval for σ, then determine whether the quality-control supervisor is in trouble.

____

7–24. ***Body Mass Index*** Refer to Data Set 1 in Appendix B.

a. Construct a 99% confidence interval estimate of the standard deviation of body mass indexes for men.

____

b. Construct a 99% confidence interval estimate of the standard deviation of body mass indexes for women

____

c. Compare and interpret the results.

____

____

7–25. ***Sample Size for Proportion*** Many states are carefully considering steps that would help them collect sales taxes on items purchased through the Internet. How many randomly selected sales transactions must be surveyed to determine the percentage that transpired over the Internet? Assume that we want to be 99% confident that the sample percentage is within two percentage points of the true population percentage for all sales transactions. __________

7–26. ***Sample Size for Proportion*** As a manufacturer of golf equipment, the Spalding Corporation wants to estimate the proportion of golfers who are left-handed. (The company can use this information in planning for the number of right-handed and left-handed sets of golf clubs to make.) How many golfers must be surveyed if we want 99% confidence that the sample proportion has a margin of error of 0.025?

a. Assume that there is no available information that could be used as an estimate of $\hat{p}$.__________

b. Assume that we have an estimate of $\hat{p}$ found from a previous study that suggests that 15% of golfers are left-handed (based on a *USA Today* report).__________

c. Assume that instead of using randomly selected golfers, the sample data are obtained by asking TV viewers of the golfing channel to call an "800" phone number to report whether they are left–handed or right–handed. How are the results affected?

____

7–27. ***Sample Size for Proportion*** You have been hired by the Ford Motor Company to do market research, and you must estimate the percentage of households in which a vehicle is owned. How many households must you survey if you want to be 94% confident that your sample percentage has a margin of error of three percentage points?

a. Assume that a previous study suggested that vehicles are owned in 86% of households.__________

b. Assume that there is no available information that can be used to estimate the percentage of households in which a vehicle is owned.__________

c. Assume that instead of using randomly selected households, the sample data are obtained by asking readers of the *Washington Post* newspaper to mail in a survey form. How are the results affected?

____

7–28. ***Sample Size for Proportion*** Concerned about campus safety, college officials want to estimate the percentage of students who carry a gun, knife, or other such weapon. How many randomly selected students must be surveyed in order to be 95% confident that the sample percentage has a margin of error of three percentage points?

a. Assume that another study indicated that 7% of college students carry weapons (based on a study by Cornell University).__________

b. Assume that there is no available information that can be used to estimate the percentage of college students carrying weapons.__________

7–29. ***Sample Size for Mean*** The Weschler IQ test is designed so that the mean is 100 and the standard deviation is 15 for the population of normal adults. Find the sample size necessary to estimate the mean IQ score of statistics students. We want to be 95% confident that our sample mean is within 2 IQ points of the true mean. The mean for this population is clearly greater than 100. The standard deviation for this population is probably less than 15 because it is a group with less variation than a group randomly selected from the general population; therefore, if we use $\sigma = 15$, we are being conservative by using a value that will make the sample size at least as large as necessary. Assume then that $\sigma = 15$ and determine the required sample size.__________

7–30. ***Sample Size for Mean*** The Tyco Video Game Corporation finds that it is losing income because of slugs used in its video games. The machines must be adjusted to accept coins only if they fall within set limits. In order to set those limits, the mean weight of quarters in circulation must be estimated. A sample of quarters will be weighed in order to determine the mean. How many quarters must we randomly select and weigh if we want to be 99% confident that the sample mean is within 0.025 g of the true population mean for all quarters? Estimate the standard deviation of the weights of all quarters by using the sample of weights (post-1964 in column 7) in Data Set 14 of Appendix B in the textbook. Standard deviation: _______Required sample size: _____

7–31. ***Sample Size for Mean*** An economist wants to estimate the mean income for the first year of work for college graduates who have had the profound wisdom to take a statistics course. How many such incomes must be found if we want to be 95% confident that the sample mean is within \$500 of the true population mean? Assume that a previous study has revealed that for such incomes, $\sigma = \$6250$.__________

7–32. ***Sample Size for Mean*** Nielsen Media Research wants to estimate the mean amount of time (in minutes) that full-time college students spend watching television each weekday. Find the sample size necessary to estimate that mean with a 15 minute margin of error. Assume that a 96% confidence level is desired. Also assume that a pilot study showed that the standard deviation is estimated to be 112.2 min.__________

7–33. ***Sample Size for Variation*** In each of the following, assume that each sample is a simple random sample obtained from a normally distributed population.

a. Find the minimum sample size needed to be 95% confident that the sample standard deviation s is within 10% of σ.__________

b. Find the minimum sample size needed to be 95% confident that the sample standard deviation s is within 30% of σ.__________

c. Find the minimum sample size needed to be 99% confident that the sample variance is within 1% of the population variance. Is such a sample size practical in most cases?__________

d. Find the minimum sample size needed to be 95% confident that the sample variance is within 20% of the population variance.__________

7-34. *Bootstrap Resampling* The sample values 2.9, 564.2, 1.4, 4.7, 67.6, 4.8, 51.3, 3.6, 18.0, and 3.6 are randomly selected from a population with a distribution that is far from normal. Use bootstrap resampling to construct a 95% confidence interval estimate of μ, and use bootstrap resampling to construct a 95% confidence interval estimate of σ. Enter the results here.

______________________________ ______________________________

Hypothesis Testing

8-1 Testing Hypotheses About a Proportion p

8-2 Testing Hypotheses About a Mean μ

8-3 Testing Hypotheses About σ or σ^2

8-4 Hypothesis Testing with Simulations

STATDISK is designed for conducting a variety of hypothesis tests included in the textbook. If you click on the main menu item of **Analysis**, then select the subdirectory item of **Hypothesis Testing**, you will now see the following menu of choices.

Proportion - One Sample

Proportion - Two Samples

St. Dev. - One Sample

St. Dev. - Two Samples

Mean - One Sample

Mean - Two Independent Samples

Mean - Matched Pairs

Among the items in the above list, three include a reference to *one sample*, and they involve hypothesis tests with claims made about a single population, as discussed in Chapter 8 of the textbook. The other items involve *two* sets of sample data as described in Chapter 9 of the textbook. This chapter will consider hypothesis testing involving only one sample. The claim may be made about the proportion of a single population, or the mean of a single population, or the standard deviation or variance of a single population. In Chapter 9 we consider claims involving two samples.

We use hypothesis testing when we want to test some claim made about a particular characteristic of some population. In addition to the claim itself, we also need sample data and a significance level.

8-1 Testing Hypotheses About a Proportion p

Section 8-2 of the textbook introduces general concepts and terminology associated with hypothesis testing, and Section 8-3 introduces procedures for testing claims about the proportion of a single population. The STATDISK procedure for testing claims about a population proportion is quite easy. In addition to having a claim to be tested, STATDISK also requires a significance level, the sample size n, and the number of successes x. In Section 7-1 of this manual/workbook, we briefly discussed one particular difficulty that arises when the available information provides the sample size n and the sample proportion $\hat{p}$ instead of the number of successes x. We provided this procedure for determining the number of successes x.

To find the number of successes x from the sample proportion and sample size:

Calculate $x = \hat{p}n$, then round the result to the nearest whole number.

Once the claim has been identified and the sample values of n and x have been determined, we can proceed to use STATDISK.

STATDISK Procedure for Testing Claims about *p*

1. Select **Analysis** from the main menu.

2. Select **Hypothesis Testing** from the subdirectory.

3. Select **Proportion - One Sample**.

4. Make these entries in the dialog box.

    -Select the format of the claim that is being tested. The default will appear as

    1) Pop. Proportion = Claimed Proportion

    and that can be changed to any of the other possibilities by using the mouse to scroll through the 6 options. Click on the box to make the other options appear, then click on the format of the claim being tested.

    -Enter a significance level, such as 0.05 or 0.01.

    -Enter the *claimed* value of the population proportion.

Consider this example: Among 703 surveyed workers, 61% said that they got their jobs through networking. Use those sample results with a 0.05 significance level to test the claim that the majority (more than 50%) of workers get their jobs through networking.

The statement that the majority got their jobs through networking is represented as $p > 0.5$. The sample data are summarized as $n = 703$ and $x = 61\%$ of $703 = 429$ (rounded to a whole number). Using STATDISK with these values, we obtain the display shown on the next page.

Important elements of the STATDISK display include the *P*-value of 0.0000, the test statistic of $z = 5.8459$, and critical value of 1.6449. Having the *P*-value and critical values available, we can use either the traditional method of testing hypotheses or the *P*-value method. For this example, we know that the *P*-value is less than the significance level, so we reject the null hypothesis of $p = 0.5$. If we were to use the traditional approach, we see that the test statistic of $z = 5.8459$ exceeds the right-tailed critical value of $z = 1.6449$, so we reject the null hypothesis of $p = 0.5$. There is sufficient evidence to support the claim that the majority of workers get their jobs through networking.

Note also that the display includes a 90% confidence interval of $0.580 < p < 0.640$ (rounded). The following points are important for interpreting this confidence interval.

1. Because the hypothesis test is right–tailed, the 0.05 significance level corresponds to a 90% confidence interval.
2. The confidence interval of $0.580 < p < 0.640$ (rounded) suggests that the population proportion p is greater than 0.5, so there is sufficient evidence to support the claim that the majority of workers get their jobs through networking.
3. The textbook notes that both the traditional method and *P*–value method use the same standard deviation based on the *claimed proportion p*, but the confidence interval uses an estimated standard deviation based on the *sample proportion* $\hat{p}$.

   Consequently, it is possible that in some cases, the traditional and *P*–value methods of testing a claim about a proportion might yield a different conclusion than the confidence interval method.

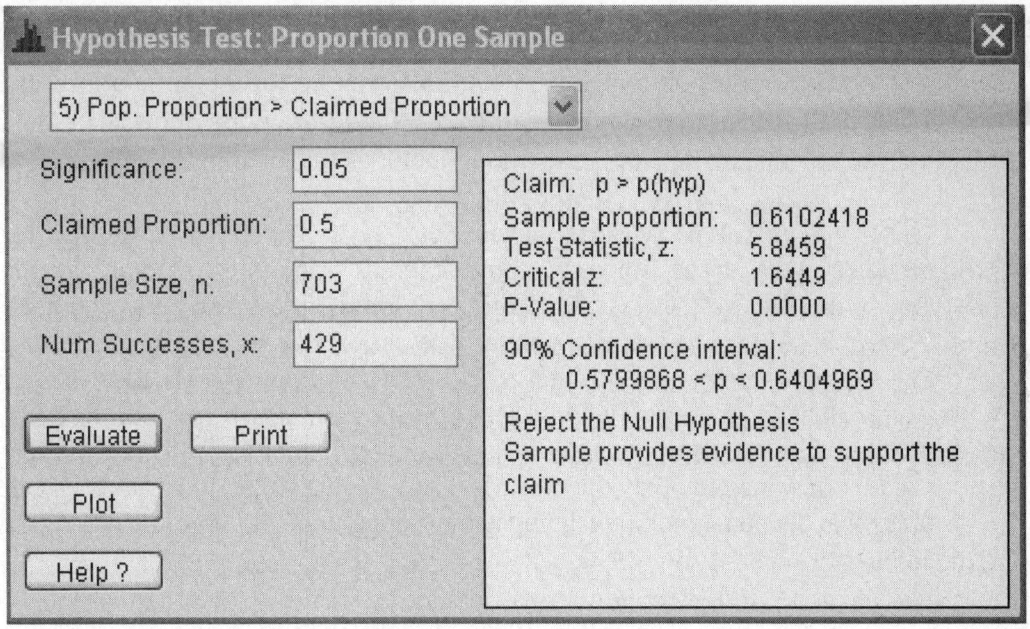

Hypothesis Test: Proportion One Sample

5) Pop. Proportion > Claimed Proportion

Significance: 0.05

Claimed Proportion: 0.5

Sample Size, n: 703

Num Successes, x: 429

Evaluate Print

Plot

Help ?

Claim: p > p(hyp)
Sample proportion: 0.6102418
Test Statistic, z: 5.8459
Critical z: 1.6449
P-Value: 0.0000

90% Confidence interval:
0.5799868 < p < 0.6404969

Reject the Null Hypothesis
Sample provides evidence to support the claim

8–2 Testing Hypotheses About a Mean μ

Sections 8–4 and 8–5 of the textbook describe methods for testing claims about a population mean μ. The textbook explains that there are different procedures, depending on the size of the sample, the nature of the population distribution, and whether the population standard deviation σ is known. The textbook stresses the importance of selecting the correct distribution (normal or t). The criteria are summarized in the table below.

Choosing between z and t

Method	Conditions
Use normal (z) distribution.	**σ known and normally distributed population** *or* **σ known and $n > 30$**
Use t distribution.	**σ not known and normally distributed population** *or* **σ not known and $n > 30$**
Use a nonparametric method or bootstrapping.	**Population is not normally distributed and $n \leq 30$**

STATDISK greatly simplifies the process of choosing between the normal and t distributions because it is programmed to make the correct choice, depending on the information that is supplied. [One exception: Like other statistics software packages, STATDISK's hypothesis testing modules are not programmed to check for normality of the population, so you should not use STATDISK if the sample size is small ($n \leq 30$) and the population has a distribution that is very non-normal.] Section 8–5 of the textbook includes this example:

> **Quality Control of M&Ms** Data Set 13 in Appendix B (in the textbook) includes weights of 13 red M&M candies randomly selected from a bag containing 465 M&Ms. Those weights (in grams) are listed below, and they have a mean of $\bar{x} = 0.8635$ and a standard deviation of $s = 0.0576$ g. The bag states that the net weight of the contents is 396.9 g, so the M&Ms must have a mean weight that is at least 396.9/465 = 0.8535 g in order to provide the amount claimed. Use the sample data with a 0.05 significance level to test the claim of a production manager that the M&Ms have a mean that is actually greater than 0.8535 g, so consumers are being given *more* than the amount indicated on the label.
>
> 0.751 0.841 0.856 0.799 0.966 0.859 0.857
>
> 0.942 0.873 0.809 0.890 0.878 0.905

In Section 1-2 of this manual/workbook we presented the method for entering data into STATDISK, and in Section 3-1 we described the procedure for obtaining descriptive statistics. If we enter the weights listed above, we will obtain these descriptive statistics:

$$n = 13 \qquad \bar{x} = 0.8635 \qquad s = 0.0576$$

STATDISK Procedure for Hypothesis Tests about a Mean

1. Select the main menu item of **Analysis**.

2. Select **Hypothesis Testing** from the subdirectory.

3. Select the first item of **Mean - One Sample.**

4. You will now see a dialog box for entry of the claim, significance level, value of the claimed mean, and entry of the sample statistics.

   -Select the format of the claim that is being tested. The default will appear as

   1) Pop. Mean = Claimed Mean

   and that can be changed to any of the other possibilities by using the mouse to scroll through the 6 options. Click on the box to make the other options appear, then click on the format of the claim being tested.

   -Enter a significance level, such as 0.05 or 0.01.

   -Enter the *claimed* value of the population mean.

   -Enter the *population* standard deviation σ if it is known. If σ is not known (as is usually the case), ignore that box and leave it empty. (*Caution*: Be careful to avoid the mistake of incorrectly entering the *sample* standard deviation in the box for the *population* standard deviation.)

   -Enter the sample size n, sample mean $\bar{x}$, and the sample standard deviation s.

   -Click the **Evaluate** button to get the test results.

   -Click the **Plot** button to get a graph that shows the test statistic and critical values.

Shown below is the STATDISK display that results from a test of the alternative hypothesis H_1: $\mu > 0.8535$ with the sample statistics of $n = 13$, $\bar{x} = 0.8635$, and $s = 0.0576$. The important elements of the display are the *P*-value of 0.2715, the test statistic of $t = 0.6260$, and the critical value of $t = 1.7823$. The *P*-value of 0.2715 is greater than the significance level of 0.05, so we fail to reject the null hypothesis of $\mu = 0.8535$ and we do not support the alternative hypothesis that $\mu > 0.8535$. Note that the display includes the conclusion of failing to reject the null hypothesis, along with the conclusion that the "sample does not provide enough evidence to support the claim," which is $\mu > 0.8535$. Note also that the display includes the 90% confidence interval of $0.8350 < \mu < 0.8920$ (rounded). The following points are important for interpreting this confidence interval.

1. Because the hypothesis test is right–tailed, the 0.05 significance level corresponds to a 90% confidence interval.

2. The confidence interval of $0.8350 < \mu < 0.8920$ (rounded) suggests that the population mean μ can be any value between 0.8350 and 0.8920, and the assumed mean of 0.8535 does fall within those limits, so there is not sufficient evidence to support the claim that $\mu > 0.8535$.

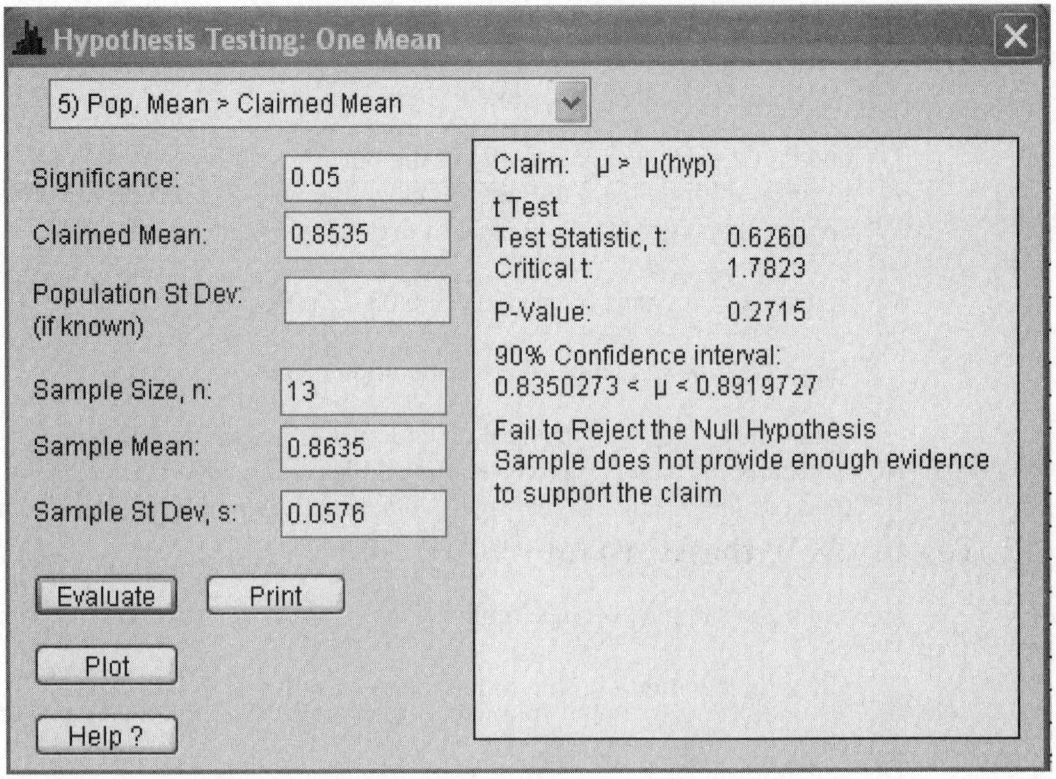

Hypothesis Testing: One Mean

5) Pop. Mean > Claimed Mean

Significance: 0.05

Claimed Mean: 0.8535

Population St Dev: (if known)

Sample Size, n: 13

Sample Mean: 0.8635

Sample St Dev, s: 0.0576

Evaluate Print

Plot

Help ?

Claim: $\mu > \mu(hyp)$

t Test

Test Statistic, t: 0.6260

Critical t: 1.7823

P-Value: 0.2715

90% Confidence interval:

$0.8350273 < \mu < 0.8919727$

Fail to Reject the Null Hypothesis

Sample does not provide enough evidence to support the claim

If you click on the **Plot** bar, you will also obtain the graph shown on the next page. Because the graph will appear in color, it will be easier to interpret on the computer monitor than the black and white reproduction shown here. The vertical line representing the test statistic will be in blue, and vertical lines representing critical values will be in red.

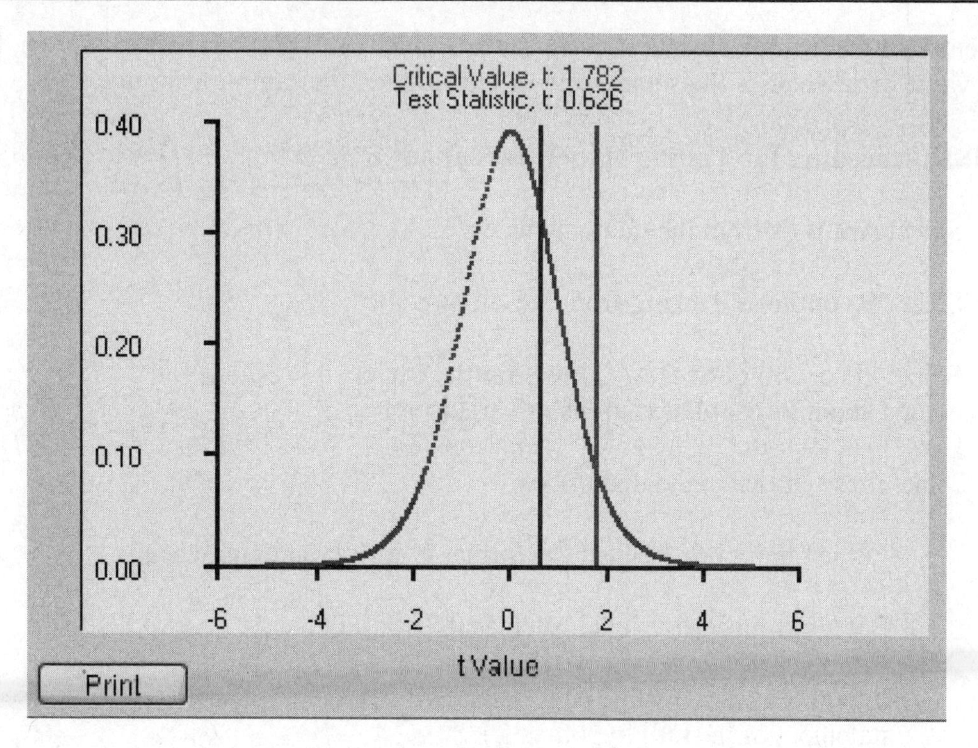

8-3 Testing Hypotheses About σ or σ^2

Caution: See Section 8-6 of the textbook, where these two important requirements are noted:

1. The samples are simple random samples. (Always remember the importance of good sampling methods.)

2. The sample values come from a population with a *normal distribution.*

The textbook makes the very important point that tests of claims about standard deviations or variances are much stricter about the requirement of a normal distribution. If the population does not have a normal distribution, then inferences about standard deviations or variances can be very misleading. *Suggestion:* Given sample data, construct a histogram and/or normal quantile plot to determine whether the assumption of a normal distribution is reasonable. If the population distribution does not appear to have a normal distribution, do not use the methods described in Section 8-6 of the textbook or this section of this manual/workbook. If the population distribution *does* appear to be normal and you want to test a claim about the population standard deviation or variance, use the STATDISK procedure given below.

Although STATDISK is designed to work only with standard deviations, claims about a population variance can be handled as well. For example, to test the claim that $\sigma^2 = 9$, take the square root of both sides of the equation, then restate the claim as $\sigma = 3$. Also, STATDISK requires entry of the sample *standard deviation s*, so if the sample variance is known, be sure to enter the value of s, which is the square root of the value of the sample variance.

STATDISK Procedure for Testing Hypotheses about σ or σ^2

1. Select **Analysis** from the main menu.

2. Select **Hypothesis Testing** from the subdirectory.

3. Select the option of **St Dev - One Sample**. (Select this option for claims about standard deviations or variances.)

4. Make these entries in the dialog box.

   -In the "claim box," select the format of the claim being tested.

   -Enter a significance level, such as 0.05 or 0.01.

   -Enter the *claimed* value of the standard deviation. (This is the value used in the statement of the null hypothesis.)

   -Enter the sample size n.

   -Enter the value of the sample standard deviation s.

This example is included in Section 8-6 of the textbook:

Quality Control The Newport Bottling Company had been manufacturing cans of cola with amounts having a standard deviation of 0.051 oz. A new bottling machine is tested, and a simple random sample of 24 cans results in the amounts (in ounces) listed below. (Those 24 amounts have a standard deviation of $s = 0.039$ oz.) Use a 0.05 significance level to test the claim that cans of cola from the new machine have amounts with a standard deviation that is less than 0.051 oz.

11.98 11.98 11.99 11.98 11.90 12.02 11.99 11.93 12.02 12.02
12.02 11.98 12.01 12.00 11.99 11.95 11.95 11.96 11.96 12.02
11.99 12.07 11.93 12.05

We want to test the claim that $\sigma < 0.051$, and we want to use a 0.05 significance level. Sample data consist of $n = 24$ and $s = 0.039$. The STATDISK display is shown below. The STATDISK display shows that the *P*-value is 0.0584. With STATDISK, it is much easier to find the *P*-value, so the *P*-value approach to hypothesis testing can be used with much greater ease. Using the *P*-value approach, we see that the *P*-value of 0.0584 is greater than the significance level of 0.05, so we fail to reject the null hypothesis that $\sigma = 15$. There is not sufficient evidence to support the claim that the population has a standard deviation less than 0.051. Because critical values are included in the STATDISK display, the traditional approach to hypothesis testing can also be used. A confidence interval for σ (denoted by SD in the display) is also displayed, along with a confidence interval for σ^2 (denoted by VAR, for variance).

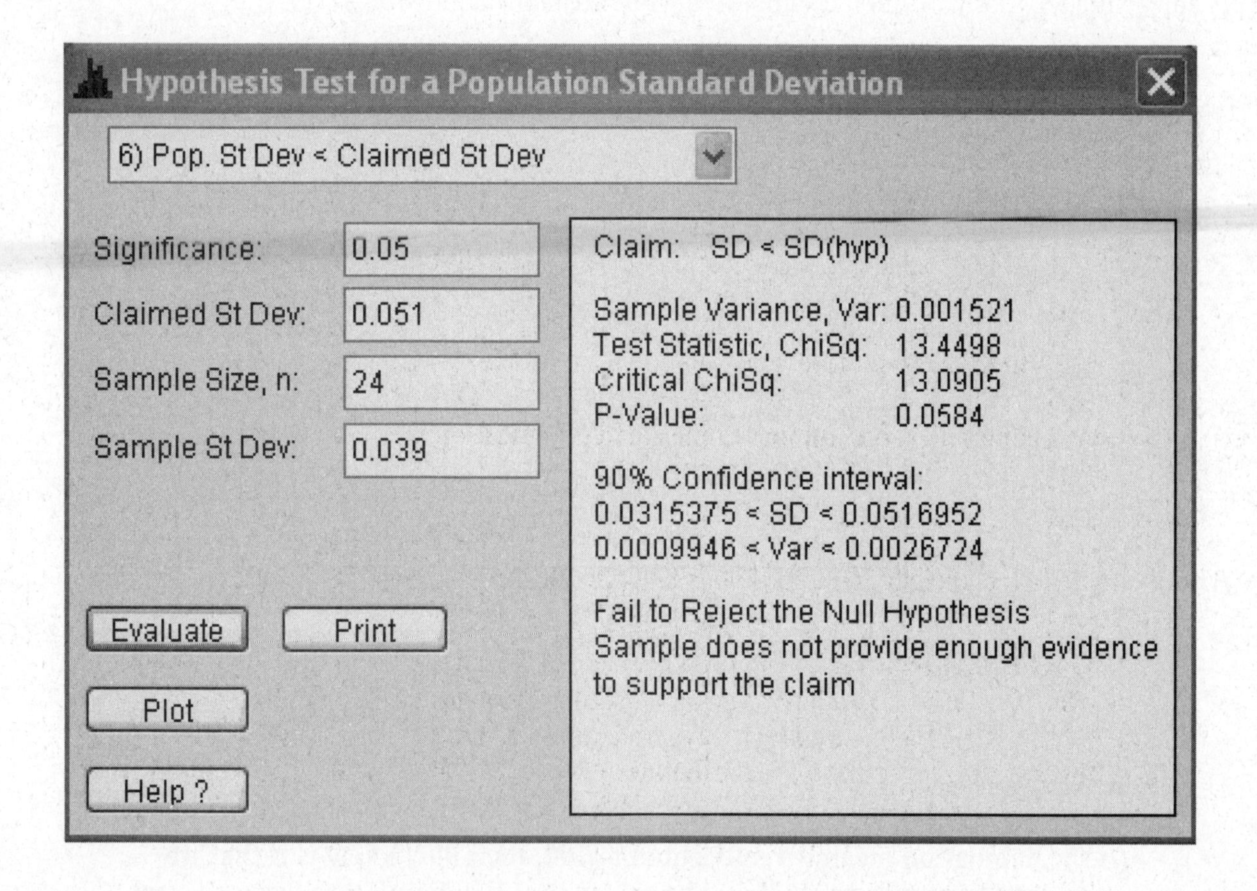

8-4 Hypothesis Testing with Simulations

Sections 8-1, 8-2, and 8-3 of this manual/workbook have all described the use of STATDISK for hypothesis tests using the traditional approach, *P*-value approach, or confidence intervals. Another very different approach is to use *simulations*. Let's illustrate the simulation technique with a simple example. Consider the claim that the following sample of IQ scores is from a population with a mean equal to 100. That is, consider the claim that $\mu = 100$.

104 105 108 122 119 → $n = 5$, $\bar{x} = 111.6$, $s = 8.3$

The following normal quantile plot shows that the sample IQ scores are from a distribution that is approximately normal, and the sample statistics are shown above.

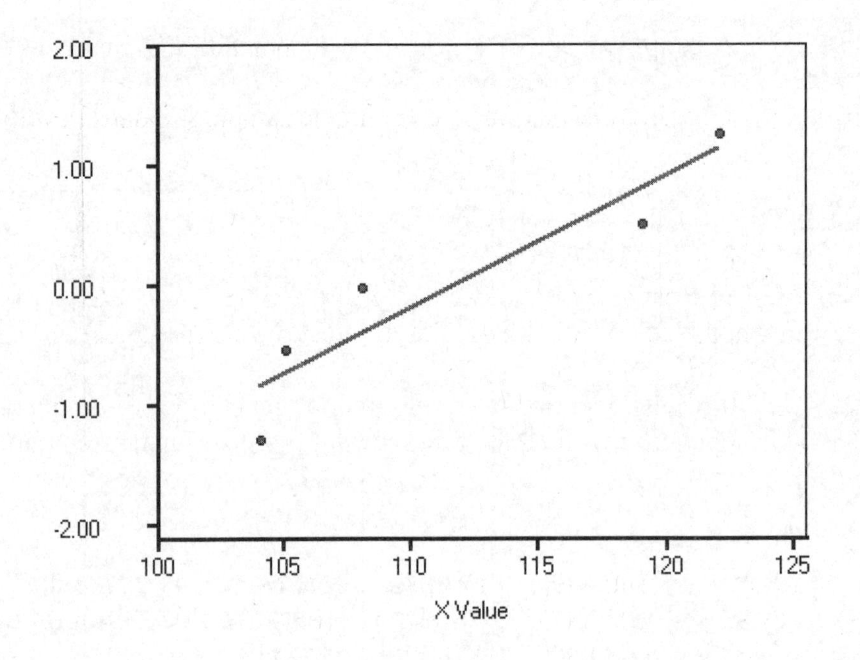

Here is the key issue:

> **If the population mean is really equal to 100, then how likely is it that we would get a sample mean like 111.6, given that the population has a normal distribution and the sample size is 5?**

If the probability of getting a sample mean such as 111.6 is very small, it appears that the sample results are not the result of chance random fluctuation. If the probability is high, then we can accept random chance as an explanation for the discrepancy between the sample mean of 111.6 and the claimed mean of 100. What we need is some way of determining the likelihood of getting a sample mean such as 111.6. That is the precise role of *P*-values in the *P*-value approach to hypothesis testing. However, there is another approach. STATDISK and many other software packages are capable of generating random results from a variety of different populations. Here is how STATDISK can be used: Determine the likelihood of getting a sample mean of 111.6 by randomly generating several different samples from a population that is normally distributed with the claimed mean of 100. For the standard deviation, we will use the best available information: the value of $s = 8.3$ obtained from the sample.

STATDISK Procedure for Testing Hypotheses with Simulations

1. Identify the values of the sample size n, the sample standard deviation s, and the claimed value of the population mean.

2. Click on **Data**.

3. Click on **Normal Generator.**

4. Generate a sample randomly selected from a population with the claimed mean. When making the required entries in the dialog box, use the *claimed* mean (not the sample mean), the sample size n, and the sample standard deviation s. After generating the sample, find its mean.

5. Continue to generate similar samples until it becomes clear that the given sample mean is or is not likely. (Here is one criterion: The given sample mean is *unlikely* if values at least as extreme occur 5% of the time or less.) If it is unlikely, reject the claimed mean. If it is likely, fail to reject the claimed mean.

For example, here are 20 results obtained from the random generation of samples of size 5 from a normally distributed population with an assumed mean of 100, a standard deviation of 8.3 (and 0 decimal places):

92.675	100.489	103.082	99.869	97.788
109.433	108.387	94.802	102.965	98.341
98.581	104.850	99.777	100.150	100.486
103.978	105.750	96.635	99.442	100.843

Examining the 20 sample means generated from a population with a mean equal to 100, we can see that *none* of them are as extreme as 111.6. Based on the generated samples, we can see that a sample mean like 111.6 is highly unlikely. This suggests that a sample mean of 111.6 is a result that is *significantly* different from the claimed mean of 100. We would feel more confident in this conclusion if we had more sample results, so we could continue to randomly generate simulated samples until we feel quite confident in our thinking that a sample mean such as 111.6 is an unusual result. We reject the null hypothesis that the sample is from a population with a mean equal to 100.

CHAPTER 8 EXPERIMENTS: Hypothesis Testing

Experiments 8–1 through 8–4 involve claims about proportions.

8–1. ***Gender Selection for Girls*** The Genetics and IVF Institute conducted a clinical trial of the XSORT method designed to increase the probability of conceiving a girl. As this book was being written, 325 babies were born to parents using the XSORT method, and 295 of them were girls. Use the sample data with a 0.01 significance level to test the claim that with this method, the probability of a baby being a girl is greater than 0.5. Does the method appear to work?

Test statistic: __________ Critical value(s): __________ *P*–value: __________

Conclusion in your own words: ____

____

8–2. ***Cloning Survey*** In a Gallup poll of 1012 randomly selected adults, 9% said that cloning of humans should be allowed. Use a 0.05 significance level to test the claim that less than 10% of all adults say that cloning of humans should be allowed. Can a newspaper run a headline that "less than 10% of all adults believe that cloning of humans should be allowed?"

Test statistic: __________ Critical value(s): __________ *P*–value: __________

Conclusion in your own words: ____

____

8–3. ***Store Checkout-Scanner Accuracy*** In a study of store checkout-scanners, 1234 items were checked and 20 of them were found to be overcharges (based on data from "UPC Scanner Pricing Systems: Are They Accurate?" by Goodstein, *Journal of Marketing,* Vol. 58). Use a 0.05 significance level to test the claim that with scanners, 1% of sales are overcharges. (Before scanners were used, the overcharge rate was estimated to be about 1%.) Based on these results, do scanners appear to help consumers avoid overcharges?

Test statistic: __________ Critical value(s): __________ *P*–value: __________

Conclusion in your own words: ____

____

8–4. ***Drug Testing of Job Applicants*** In 1990, 5.8% of job applicants who were tested for drugs failed the test. At the 0.01 significance level, test the claim that the failure rate is now lower if a simple random sample of 1520 current job applicants results in 58 failures (based on data from the American Management Association). Does the result suggest that fewer job applicants now use drugs?

Test statistic: __________ Critical value(s): __________ *P*–value: __________

Conclusion in your own words: ____

____

Experiments 8–5 through 8–12 require tests of claims about means of populations.

8–5. ***Monitoring Lead in Air*** Listed below are measured amounts of lead (in micrograms per cubic meter or $\mu g/m^3$) in the air. The Environmental Protection Agency has established an air quality standard for lead: 1.5 $\mu g/m^3$. The measurements shown below were recorded at Building 5 of the World Trade Center site on different days immediately following the destruction caused by the terrorist attacks of September 11, 2001. After the collapse of the two World Trade buildings, there was considerable concern about the quality of the air. Use a 0.05 significance level to test the claim that the sample is from a population with a mean greater than the EPA standard of 1.5 $\mu g/m^3$. Is there anything about this data set suggesting that the assumption of a normally distributed population might not be valid?

5.40 1.10 0.42 0.73 0.48 1.10

Test statistic: __________ Critical value(s): __________ *P*–value: __________

Conclusion in your own words: ____

____

8–6. ***Treating Chronic Fatigue Syndrome*** Patients with chronic fatigue syndrome were tested, then retested after being treated with fludrocortisone. Listed below are the changes in fatigue after the treatment (based on data from "The Relationship Between Neurally Mediated Hypotension and the Chronic Fatigue Syndrom" by Bou–Holaigah, Rowe, Kan, and Calkins, *Journal of the American Medical Association,* Vol. 274, No. 12). A standard scale from –7 to +7 was used, with positive values representing improvements. Use a 0.01 significance level to test the claim that the mean change is positive. Does the treatment appear to be effective?

6 5 0 5 6 7 3 3 2 6 5 5 0 6 3 4 3 7 0 4 4

Test statistic: __________ Critical value(s): __________ *P*–value: __________

Conclusion in your own words: ____

8–7. ***Olympic Winners*** Listed below are the winning times (in seconds) of men in the 100-meter dash for consecutive summer Olympic games, listed in order by row. Assuming that these results are sample data randomly selected from the population of all past and future Olympic games, test the claim that the mean time is less than 11 sec. What do you observe about the precision of the numbers? What extremely important characteristic of the data set is not considered in this hypothesis test? Do the results from the hypothesis test suggest that future winning times should be around 10.5 sec, and is such a conclusion valid?

12.0	11.0	11.0	11.2	10.8	10.8	10.8	10.6	10.8	10.3	10.3	10.3
10.4	10.5	10.2	10.0	9.95	10.14	10.06	10.25	9.99	9.92	9.96	

Test statistic: __________ Critical value(s): __________ *P*–value: __________

Conclusion in your own words: ____

____

8–8. ***Nicotine in Cigarettes*** The Carolina Tobacco Company advertised that its best-selling nonfiltered cigarettes contain at most 40 mg of nicotine, but *Consumer Advocate* magazine ran tests of 10 randomly selected cigarettes and found the amounts (in mg) shown in the accompanying list. It's a serious matter to charge that the company advertising is wrong, so the magazine editor chooses a significance level of $\alpha = 0.01$ in testing her belief that the mean nicotine content is greater than 40 mg. Using a 0.01 significance level, test the editor's belief that the mean is greater than 40 mg.

47.3 39.3 40.3 38.3 46.3 43.3 42.3 49.3 40.3 46.3

Test statistic: __________ Critical value(s): __________ *P*–value: __________

Conclusion in your own words: ____

____

8–9. ***Testing Wristwatch Accuracy*** Students of the author randomly selected 40 people and measured the accuracy of their wristwatches, with positive errors representing watches that are ahead of the correct time and negative errors representing watches that are behind the correct time. The 40 values have a mean of 117.3 sec and a standard deviation of 185.0 sec. Use a 0.01 significance level to test the claim that the population of all watches has a mean equal to 0 sec. What can be concluded about the accuracy of people's wristwatches?

Test statistic: __________ Critical value(s): __________ *P*–value: __________

Conclusion in your own words: ____

8–10. ***World's Smallest Mammal*** The world's smallest mammal is the bumblebee bat, also known as the Kitti's hog-nosed bat (or Craseonycteris thonglongyai). Such bats are roughly the size of a large bumblebee. Listed below are weights (in grams) from a sample of these bats. Test the claim that these bats come from the same population having a mean weight equal to 1.8 g.

1.7 1.6 1.5 2.0 2.3 1.6 1.6 1.8 1.5 1.7 2.2 1.4 1.6 1.6 1.6

Test statistic: __________ Critical value(s): __________ *P*–value: __________

Conclusion in your own words: ____

____

8–11. ***Conductor Life Span*** A *New York Times* article noted that the mean life span for 35 male symphony conductors was 73.4 years, in contrast to the mean of 69.5 years for males in the general population. Assuming that the 35 males have life spans with a standard deviation of 8.7 years, use a 0.05 significance level to test the claim that male symphony conductors have a mean life span that is greater than 69.5 years. Does it appear that male symphony conductors live longer than males from the general population? Why doesn't the experience of being a male symphony conductor cause men to live longer? (*Hint*: Are male symphony conductors born, or do they become conductors at a much later age?)

Test statistic: __________ Critical value(s): __________ *P*–value: __________

Conclusion in your own words: ____

____

8–12. ***Baseballs*** In previous tests, baseballs were dropped 24 ft onto a concrete surface, and they bounced an average of 92.84 in. In a test of a sample of 40 new balls, the bounce heights had a mean of 92.67 in. and a standard deviation of 1.79 in. (based on data from Brookhaven National Laboratory and *USA Today*). Use a 0.05 significance level to determine whether there is sufficient evidence to support the claim that the new balls have bounce heights with a mean different from 92.84 in. Does it appear that the new baseballs are different?

Test statistic: __________ Critical value(s): __________ *P*–value: __________

Conclusion in your own words: ____

____

Experiments 8–13 through 8–16 involve claims about a standard deviation or variance.

8–13. ***Supermodel Weights*** Use a 0.01 significance level to test the claim that weights of female supermodels vary less than the weights of women in general. The standard deviation of weights of the population of women is 29 lb. Listed below are the weights (in pounds) of nine randomly selected supermodels.

125 (Taylor) 119 (Auermann) 128 (Schiffer) 128 (MacPherson)
119 (Turlington) 127 (Hall) 105 (Moss) 123 (Mazza)
115 (Hume)

Test statistic: ________ Critical value(s): ________ *P*–value: ________

Conclusion in your own words: ____

____

8–14. ***Supermodel Heights*** Use a 0.05 significance level to test the claim that heights of female supermodels vary less than the heights of women in general. The standard deviation of heights of the population of women is 2.5 in. Listed below are the heights (in inches) of randomly selected supermodels (Taylor, Harlow, Mulder, Goff, Evangelista, Avermann, Schiffer, MacPherson, Turlington, Hall, Crawford, Campbell, Herzigova, Seymour, Banks, Moss, Mazza, Hume).

71	71	70	69	69.5	70.5	71	72	70
70	69	69.5	69	70	70	66.5	70	71

Test statistic: ________ Critical value(s): ________ *P*–value: ________

Conclusion in your own words: ____

____

8–15. ***Volumes of Pepsi*** A new production manager claims that the volumes of cans of regular Pepsi have a standard deviation less than 0.10 oz. Use a 0.05 significance level to test that claim with the sample results listed in the Cola data set in Appendix B from the textbook. What problems are caused by a mean that is not 12 oz? What problems are caused by a standard deviation that is too high?

Test statistic: ________ Critical value(s): ________ *P*–value: ________

Conclusion in your own words: ____

____

8–16. ***Systolic Blood Pressure for Women*** Systolic blood pressure results from contraction of the heart. Based on past results from the National Health Survey, it is claimed that women have systolic blood pressures with a mean and standard deviation of 130.7 and 23.4, respectively. Use the systolic blood pressures of women listed in Data Set 1 in Appendix B from the textbook and test the claim that the sample comes from a population with a standard deviation of 23.4. (The data set is already stored in STATDISK.) Enter the results below.

Test statistic: __________ Critical value(s): __________ *P*–value: __________

Conclusion in your own words: ____

____

Experiments 8-17 through 8-20 involve the simulation approach to hypothesis testing.

8–17. ***Hypothesis Testing with Simulations*** Consider a test of the claim that $\mu = 75$. The following sample results were obtained.

91 85 94 86 82

a. Generate and print a normal quantile plot. Based on the result, does this sample appear to be from a population with a normal distribution? Why or why not?

____

____

b. What does the value of the sample mean suggest about the claim that $\mu = 75$?

____

c. Generate different samples of size $n = 5$ until getting a sense for the likelihood of getting a sample mean like the one obtained. (*Hint:* See Section 8-4 in this manual/workbook.) List the sample means below.

____

____

d. What do you conclude? Why?

____

____

8–18. ***Hypothesis Testing with Simulations*** Use a simulation approach for conducting the hypothesis test described in Experiment 8–6. Describe the procedure, results, and conclusions.

____

____

____

8–19 ***Hypothesis Testing with Simulations*** Use a simulation approach for conducting the hypothesis test described in Experiment 8–7. Describe the procedure, results, and conclusions.

____

____

____

8–20. ***Hypothesis Testing with Simulations*** Use a simulation approach for conducting the hypothesis test described in Experiment 8–8. Describe the procedure, results, and conclusions.

____

____

____

8–21. ***Activities with STATDISK: P-Values*** A contractor claims that the average bi-monthly cost of electricity in a certain community is more than $250. Test the contractor's claim by using a 5% level of significance. Use the data from the data set Electricity Consumption. Assume that each of the 15 values is from a randomly selected home in the community.

Use STATDISK to find the sample mean and sample standard deviation, and round the results to the nearest dollar. Enter the results here.

Mean: __________ Standard deviation: ______________

Verify that the STATDISK graph for the hypothesis test is as shown below.

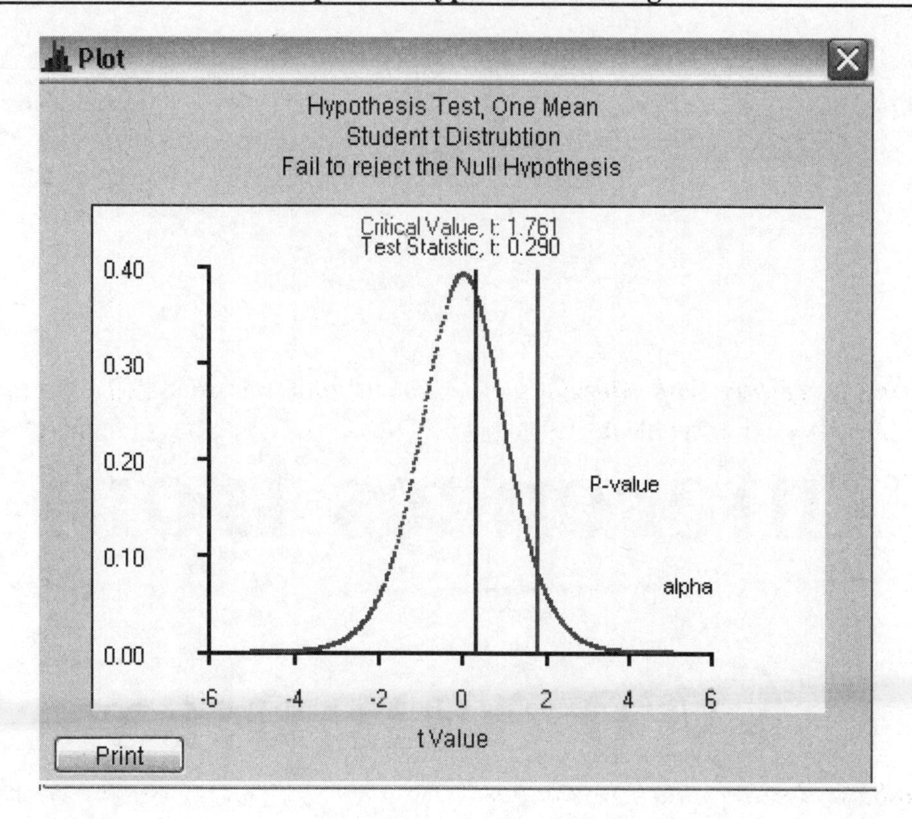

In the above graph, use a pen to drawn an arrow from "alpha" so that it points to the area representing α .

In the above graph, use a pencil to shade in the area representing the *P*-value. Also draw an arrow from "P-value" to the shaded area.

What is a *P*-value? (Describe it in your own words.)

____

____

Using the results shown in the above display, state the conclusion in your own words.

____

____

Inferences from Two Samples

9-1 Introduction

9-2 Two Proportions

9-3 Two Means: Independent Samples

9-4 Two Means: Matched Pairs

9-5 Two Variances

9-1 Introduction

As the use of technology grows in introductory statistics courses, more instructors are expanding the scope of topics covered. They often accomplish this with assignments from chapters not formally covered in class. Once students understand the basic concepts of confidence interval construction and hypothesis testing, it becomes relatively easy to use technology like STATDISK to apply that understanding to other circumstances, such as those involving two samples instead of only one.

This chapter deals with inferences based on two sets of sample data. Some of the most important applications of statistics require the methods of this chapter, such as determining whether the proportion of adverse reactions in a sample of people using a new drug is the same as the proportion of adverse reactions in a sample of people using a placebo.

Earlier versions of STATDISK were designed so that when working with two samples, you had to use the Hypothesis Testing module for hypothesis testing and confidence intervals. The new version of STATDISK has an expanded Confidence Intervals module, so you should now select Hypothesis Testing for hypothesis tests and Confidence Intervals for confidence intervals.

9-2 Two Proportions

The textbook makes the point that a comparison of two sample proportions is extremely important, because it involves situations that occur so often in real applications.

When working with two proportions, STATDISK requires that you first identify the number of successes x_1 and the sample size n_1 for the first sample, and you must also identify the values of x_2 and n_2 for the second sample. However, sample data often consist of sample proportions instead of the actual numbers of successes. Be sure to read the textbook carefully for a way to determine the number of successes. Consider the statement that "when 734 men were treated with Viagra, 16% of them experienced headaches." From that statement we can see that $n_1 = 734$ and $\hat{p}_1 = 0.16$, but the actual number of successes x_1 is not given. However, from $\hat{p}_1 = x_1/n_1$, we know that $x_1 = n_1 \cdot \hat{p}_1$ so that $x_1 = 734 \cdot 0.16 = 117.44$. Because you cannot have 117.44 men who experienced headaches, x_1 must be a whole number and we round it to 117. We now use $x_1 = 117$ in the calculations that require its value. In general, given a sample proportion and sample size, calculate the number of successes by multiplying the decimal form of the sample proportion and the sample size.

STATDISK Procedure for Testing Hypotheses About Two Proportions
To conduct hypothesis tests about two population proportions, use the following procedure.

1. For both samples, find the sample size n and the number of successes x.

2. Select **Analysis** from the main menu.

3. Select **Hypothesis Testing** from the subdirectory.

4. Select **Proportion - Two Samples**.

5. Make these entries in the dialog box:

   -In the "claim" box, select the format corresponding to the claim.

   -Enter a significance level, such as 0.05 or 0.01.

   -For each sample, enter the sample size n and the number of successes x.

6. Click on the **Evaluate** button to obtain the test results.

7. Click on the **Plot** button to obtain a graph that includes the test statistic and critical value(s).

Consider this example:

> **Is Surgery Better than Splinting?** The table below lists results from a clinical trial in which patients were treated for carpal tunnel syndrome. Use a 0.05 significance level to test the claim that the success rate with surgery is better than the success rate with splinting.

Table 9-1 **Treatments of Carpal Tunnel Syndrome**

	Surgery	Splint
Success a year after treatment	67	60
Total number treated	73	83
Success Rate	**92%**	**72%**

For notation purposes, we stipulate that sample 1 is the surgery treatment group, and sample 2 is the splint treatment group. We can summarize the sample data as follows.

Surgery Treatment	Splint Treatment
$n_1 = 73$	$n_2 = 83$
$x_1 = 67$	$x_2 = 60$
$\hat{p}_1 = \frac{x_1}{n_1} = \frac{67}{73} = 0.918$	$\hat{p}_2 = \frac{x_2}{n_2} = \frac{60}{83} = 0.723$

We can now proceed to use the above STATDISK procedure. After selecting **Analysis, Hypothesis Testing, Proportion - Two Samples**, we make the required entries in the dialog box as shown below. Note the form of the claim used to test the claim that the first population has a proportion that is *greater than* the proportion for the second population. The STATDISK results include the pooled proportion, test statistic, critical value, and *P*–value. Because the *P*-value of

0.0009 is less than the significance level of $\alpha = 0.05$, we reject the null hypothesis of $p_1 = p_2$. We conclude that there is sufficient evidence to support the claim that the proportion of successes with surgery is greater than that for splinting.

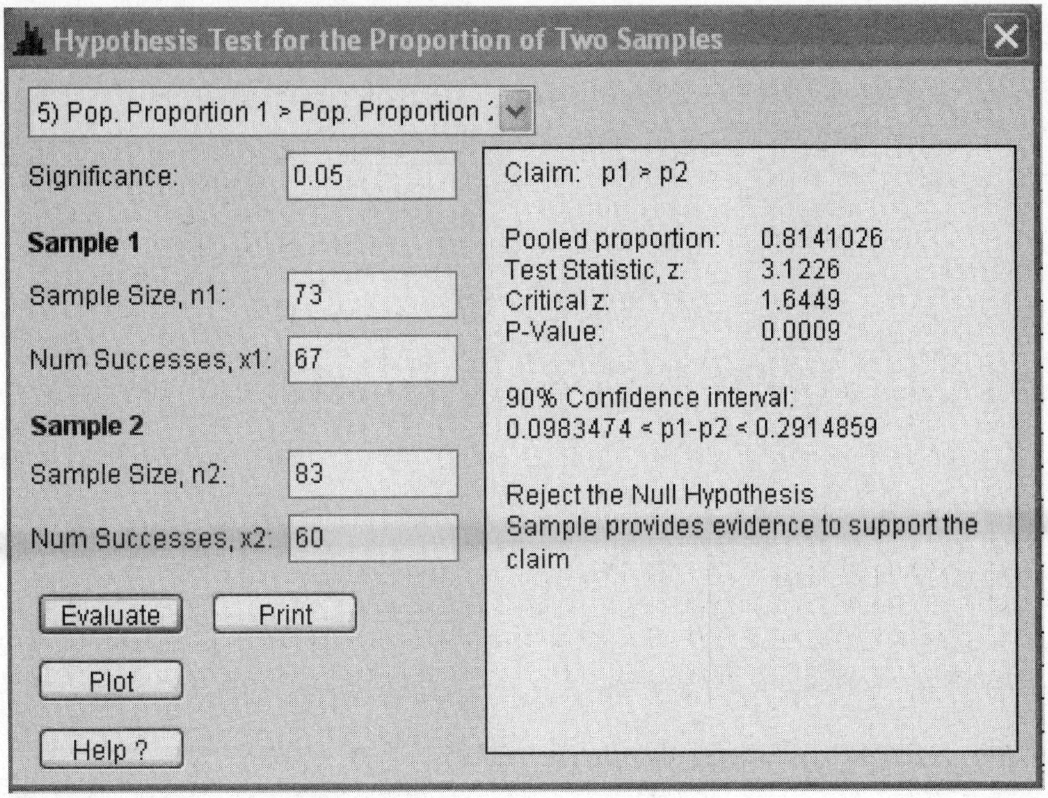

Hypothesis Test for the Proportion of Two Samples

5) Pop. Proportion 1 > Pop. Proportion 2

Significance: 0.05

Sample 1

Sample Size, n1: 73

Num Successes, x1: 67

Sample 2

Sample Size, n2: 83

Num Successes, x2: 60

Evaluate | Print | Plot | Help ?

Claim: p1 > p2

Pooled proportion: 0.8141026
Test Statistic, z: 3.1226
Critical z: 1.6449
P-Value: 0.0009

90% Confidence interval:
0.0983474 < p1-p2 < 0.2914859

Reject the Null Hypothesis
Sample provides evidence to support the claim

STATDISK Procedure for a Confidence Interval Estimate of the Difference Between Two Proportions

1. For each of the two samples, find the sample size n and the number of successes x.
2. Select **Analysis** from the main menu.
3. Select **Confidence Intervals** from the subdirectory.
4. Select **Proportion - Two Samples**.
5. Enter the confidence level.
6. Enter the sample size and number of successes for each of the two samples.
7. Click on the **Evaluate** button.

Shown below is a STATDISK display corresponding to the surgery/split sample data in the preceding table. The confidence level of 0.90 was chosen so that the confidence interval corresponds to the one-sided hypothesis test conducted with a 0.05 significance level. Note that the confidence interval limits do not include zero, suggesting that there is a significant difference between the two proportions.

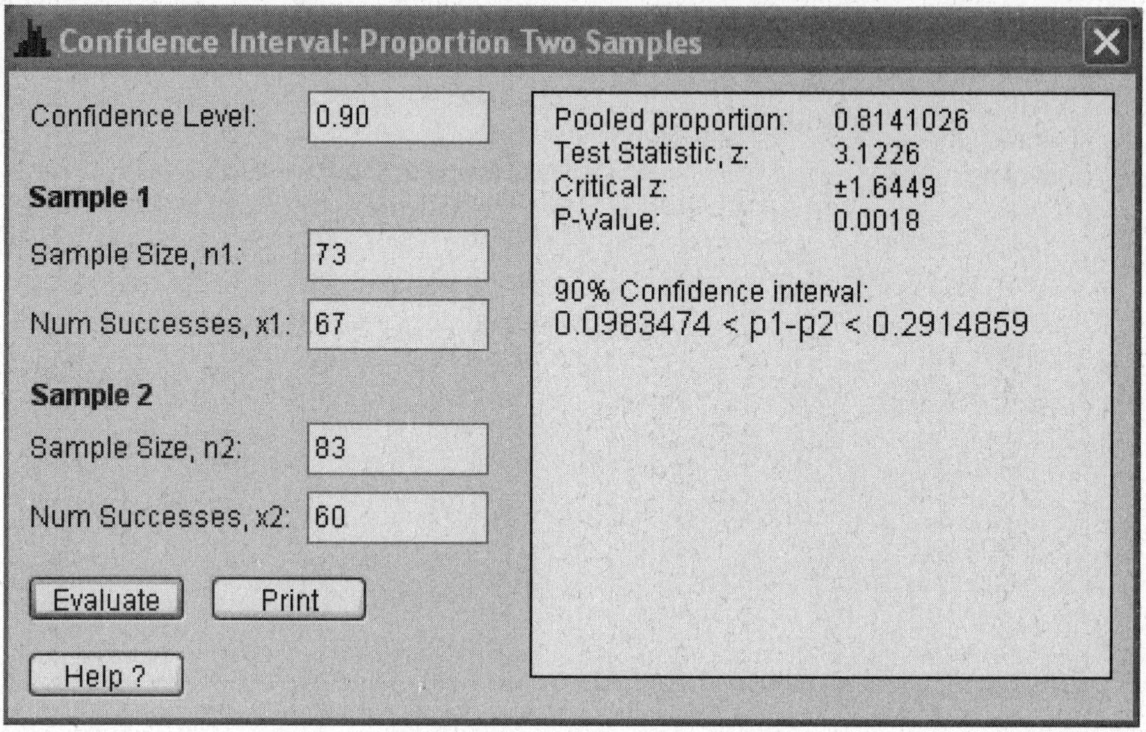

Confidence Interval: Proportion Two Samples

Confidence Level: 0.90

Sample 1

Sample Size, n1: 73

Num Successes, x1: 67

Sample 2

Sample Size, n2: 83

Num Successes, x2: 60

Evaluate Print

Help ?

Pooled proportion: 0.8141026
Test Statistic, z: 3.1226
Critical z: ±1.6449
P-Value: 0.0018

90% Confidence interval:
0.0983474 < p1-p2 < 0.2914859

9-2 Two Means: Independent Samples

Section 9–3 in the textbook is devoted to inferences about means based on two independent samples. The textbook notes that two samples are **independent** if the sample values selected from one population are not related to or somehow paired or matched with the sample values selected from the other population. If there is some relationship so that each value in one sample is paired with a corresponding value in the other sample, the samples are **dependent.** Dependent samples are often referred to as **matched pairs,** or **paired samples.**

When testing a claim about the means of two independent samples, or when constructing a confidence interval estimate of the difference between the means of two independent samples, Section 9–3 in the textbook describes procedures based on the assumptions that the two population standard deviations σ_1 and σ_2 are not known, and there is no assumption that $\sigma_1 = \sigma_2$. Section 9–3 focuses on the first of these three cases, but the other two cases are discussed briefly:

1. σ_1 and σ_2 are not known and are not assumed to be equal.
2. σ_1 and σ_2 are known.
3. It is assumed that $\sigma_1 = \sigma_2$.

Although the first of these three cases is the main focus of Section 9–3 in the textbook, STATDISK allows us to work with all three cases.

STATDISK Procedure for Tests of Hypotheses about Two Means: Independent Samples

1. For each of the two samples, identify the sample size, sample mean, and sample standard deviation. That is, find the values of n_1, $\bar{x}_1$, s_1, n_2, $\bar{x}_2$, and s_2. (If necessary, use STATDISK's Descriptive Statistics module to find the required sample statistics.)

2. Select **Analysis** from the main menu.

3. Select **Hypothesis Testing** from the submenu.

4. Choose the option of **Mean - Two Independent Samples**.

5. Enter the required values in the dialog box. Be particularly careful with these items:
   -Claim: Be sure to select the form of the claim to be tested.
   -Avoid confusion between the *sample* standard deviation and the *population* standard deviation. Values of the population standard deviation are rarely know, so those boxes are usually left blank.

6. If σ_1 and σ_2 are not known and there is no sound reason to assume that $\sigma_1 = \sigma_2$, click on the first bar for NO POOL (which means that the sample variances will not be pooled as described in Section 9–3 of the textbook) as shown below. [The option of "Eq. vars: POOL" is used when there is a sound reason to assume that $\sigma_1 = \sigma_2$, so that the sample variances will be pooled to form an estimate of the population variance. The option of "Prelim F–test" is not recommended, but it conducts a preliminary F test of the null hypothesis that $\sigma_1 = \sigma_2$ and, based on the results, proceeds with one of these two cases: (1) Do not assume that $\sigma_1 = \sigma_2$ and do not pool the sample variances; (2) Assume that $\sigma_1 = \sigma_2$ and pool the sample variances.]

   Method of analysis
   - (•) Not eq. vars: No Pool
   - ( ) Eq vars: POOL
   - ( ) Prelim F-Test

7. Click **Evaluate.**

Consider the following example.

Discrimination Based on Age The Revenue Commissioners in Ireland conducted a contest for promotion. The ages of the unsuccessful and successful applicants are given below (based on data from "Debating the Use of Statistical Evidence in Allegations of Age Discrimination" by Barry and Boland, *The American Statistician,* Vol. 58, No. 2). Some of the applicants who were unsuccessful in getting the promotion charged that the competition involved discrimination based on age. Treat the data as samples from larger populations and use a 0.05 significance level to test the claim that the unsuccessful applicants are from a population with a greater mean age than the mean age of successful applicants.

Ages of Unsuccessful Applicants	Ages of Successful Applicants
34 37 37 38 41 42 43 44 44 45	27 33 36 37 38 38 39 42 42 43
45 45 46 48 49 53 53 54 54 55	43 44 44 44 45 45 45 45 46 46
56 57 60	47 47 48 48 49 49 51 51 52 54

Because the samples are small, we must verify the requirement that the are from normally distributed populations. We could use STATDISK to generate histograms and/or normal quantile plots to see that the samples do appear to come from populations with normal distributions.

The STATDISK results are shown below. Because the *P*-value of 0.0504 is greater than the significance level, we fail to reject $\mu_1 = \mu_2$ and conclude that there is not sufficient evidence to support the claim that the unsuccessful applicants have a higher mean age.

Note that STATDISK uses the complicated Formula 9-1 for computing the number of degrees of freedom, whereas the textbook uses "the smaller of $n_1 - 1$ and $n_2 - 1$." Consequently, STATDISK results will be somewhat different than those obtained by letting the number of degrees of freedom equal to the smaller of $n_1 - 1$ and $n_2 - 1$. By using Formula 9-1, STATDISK provides better results.

```
Not eq. vars: No Pool (and df calculated with Formula 9-1)
Claim:  µ1> µ2

Test Statistic, t:       1.6777
Critical t:              1.681922
P-Value:                 0.0504
Degrees of freedom:  42.0315

90% Confidence interval:
-0.0078021 < µ1-µ2 < 6.207802

Fail to Reject the Null Hypothesis
Sample does not provide enough evidence to support the
claim
```

STATDISK Procedure for Confidence Interval Estimates of the Difference Between Two Means: Independent Samples

1. For each of the two samples, identify the sample size, sample mean, and sample standard deviation. That is, find the values of n_1, $\bar{x}_1$, s_1, n_2, $\bar{x}_2$, and s_2. (If necessary, use STATDISK's Descriptive Statistics module to find the required sample statistics.)

2. Select **Analysis** from the main menu.

3. Select **Confidence Intervals** from the submenu.

4. Choose the option of **Mean - Two Independent Samples**.

5. Enter a significance level.

6. Enter the statistics for each of the two samples.

7. If σ_1 and σ_2 are not known and there is no sound reason to assume that $\sigma_1 = \sigma_2$, click on the first bar for NO POOL (which means that the sample variances will not be pooled as described in Section 9–3 of the textbook) as shown below. [The option of "Eq. vars: POOL" is used when there is a sound reason to assume that $\sigma_1 = \sigma_2$, so that the sample variances will be pooled to form an estimate of the population variance. The option of "Prelim F–test" is not recommended, but it conducts a preliminary F test of the null hypothesis that $\sigma_1 = \sigma_2$ and, based on the results, proceeds with one of these two cases: (1) Do not assume that $\sigma_1 = \sigma_2$ and do not pool the sample variances; (2) Assume that $\sigma_1 = \sigma_2$ and pool the sample variances.]

8. Click **Evaluate.**

If this procedure is used with the sample data from the preceding example, the 90% confidence interval is included in the following STATDISK display. Because the confidence interval limits do include zero, there does not appear to be significant difference between the two means.

```
Not eq. vars: No Pool (and df calculated with Formula 9-1)

Test Statistic, t:      1.6777
Critical t:             ±1.681922
P-Value:                0.1008
Degrees of freedom:     42.0315

90% Confidence interval:
-0.0078021 < μ1-μ2 < 6.207802
```

9-4 Two Means: Matched Pairs

Section 9-4 of the textbook describes methods for testing hypotheses and constructing confidence interval estimates of the differences between samples consisting of *matched pairs*. Note that there is a requirement that the number of matched pairs of sample data must be large ($n > 30$) or the pairs of values have differences that are from a population having a distribution that is approximately normal.

STATDISK Procedure for Testing Hypothesis About the Mean of the Differences from Matched Pairs.

1. Get the paired data into two columns of the Statdisk data window. Either manually enter the data (if the lists are not very long) or open existing data sets. (If the number of matched pairs is small ($n \leq 30$), use a histogram or normal quantile plot to verify that the differences appear to come from a population with a normal distribution.)

2. Select **Analysis** from the main menu.

3. Select **Hypothesis Testing** from the subdirectory.

4. Select **Mean - Matched Pairs.**

5. Make these entries and selections in the dialog box:

   -Select the appropriate format in the "claim" box. The default is Pop. Mean of Difference = 0, which is equivalent to $\mu_d = 0$. Scroll through the other 5 possibilities and select the format corresponding to the claim being tested.

   -Enter a significance level, such as 0.05 or 0.01.

   -Select the columns of the data window to be used for the calculations.

   -Click the **Evaluate** button.

Consider this example:

Are Forecast Temperatures Accurate? Table 9-2 consists of five actual low temperatures and the corresponding low temperatures that were predicted five days earlier. The data consist of matched pairs, because each pair of values represents the same day. Use a 0.05 significance level to test the claim that there is a difference between the actual low temperatures and the low temperatures that were forecast five days earlier.

Table 9–2 Actual and Forecast Temperatures

Actual Low	54	54	55	60	64
Low Forecast Five Days Earlier	56	57	59	56	64
Difference d = Actual – Predicted	–2	–3	–4	4	0

Because the sample is small, we should verify that the sample differences appear to come from a population with a normal distribution. A histogram is not helpful with only 5 sample values, but a normal quantile plot does suggest that the differences appear to be from a normally distributed population. We can therefore proceed with the hypothesis test.

Using the STATDISK procedure for matched pairs and the sample data in Table 9–2, we get the following display when testing the claim that there is a difference between the means of the actual low temperatures and the low temperatures forecast five days earlier. The *P*–value of 0.5185 suggests that the difference is not significant, so there is not sufficient evidence to support the claim of a difference.

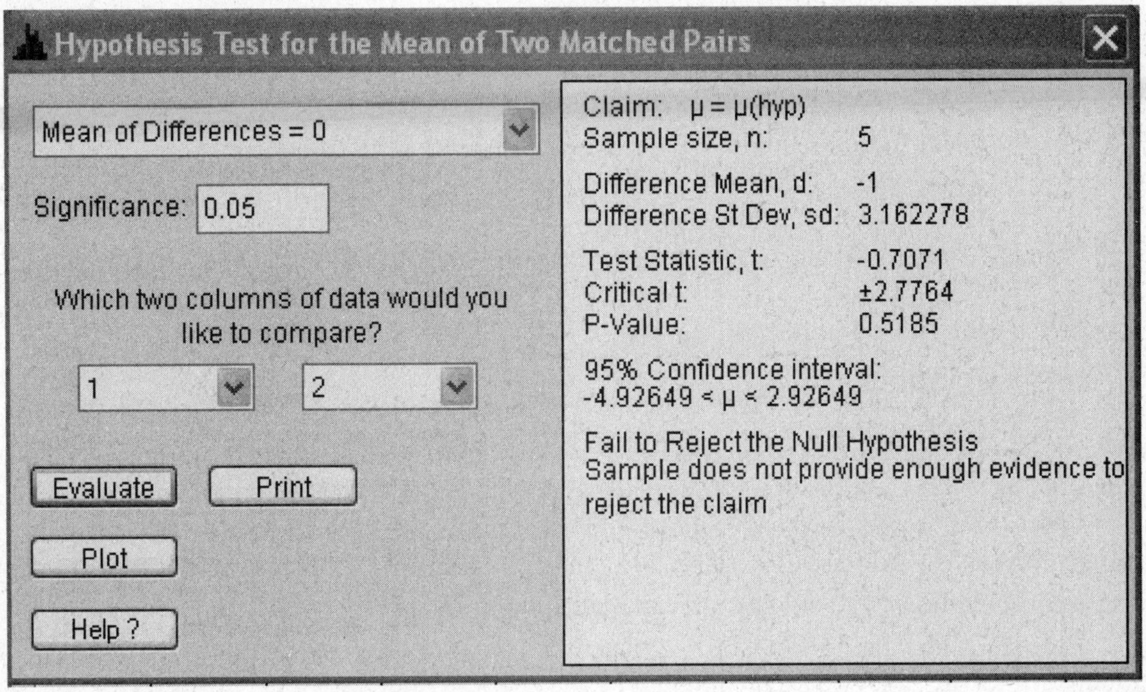

Hypothesis Test for the Mean of Two Matched Pairs

Mean of Differences = 0

Significance: 0.05

Which two columns of data would you like to compare?

1 2

Evaluate Print

Plot

Help ?

Claim: μ = μ(hyp)
Sample size, n: 5

Difference Mean, d: -1
Difference St Dev, sd: 3.162278

Test Statistic, t: -0.7071
Critical t: ±2.7764
P-Value: 0.5185

95% Confidence interval:
$-4.92649 < \mu < 2.92649$

Fail to Reject the Null Hypothesis
Sample does not provide enough evidence to reject the claim

STATDISK Procedure for a Confidence Interval Estimate of the Mean of the Differences from Matched Pairs.

1. Get the paired data into two columns of the Statdisk data window. Either manually enter the data (if the lists are not very long) or open existing data sets. (If the number of matched pairs is small ($n \leq 30$), use a histogram or normal quantile plot to verify that the differences appear to come from a population with a normal distribution.)

2. Select **Analysis** from the main menu.

3. Select **Confidence Intervals** from the subdirectory.

4. Select **Mean - Matched Pairs.**

5. Enter a confidence level and select the columns containing the sample data.

6. Click the **Evaluate** button.

Shown below is the STATDISK display resulting from the data in Table 9-2. Note that the confidence interval limits do contain 0, indicating that the true value of μ_d is not significantly different from 0. We cannot conclude that there is a significant difference between the actual and forecast low temperatures

```
Sample size, n:           5

Difference Mean, d:      -1
Difference St Dev, sd:  3.162278

Test Statistic, t:       -0.7071
Critical t:              ±2.7764
P-Value:                 0.5185

95% Confidence interval:
-4.92649 < μ < 2.92649
```

9-5 Two Variances

Section 9-5 in the textbook describes the use of the *F* distribution in testing a claim that two populations have the same variance (or standard deviation). STATDISK has a module for such tests. Section 9-5 in the textbook focuses on hypothesis tests, and confidence intervals are included only as an exercise (Exercise 24), but STATDISK can be used to find confidence intervals for the ratio σ_1^2 / σ_2^2 and also for the ratio σ_1 / σ_2.

Before using the STATDISK procedures for testing hypotheses or constructing confidence intervals, the requirements should be verified. When making inferences about two standard deviations or variances, we require that the two samples are independent. Also, the two populations must be *normally distributed.* (Verification of normality is important because the methods of this section are not robust, meaning that they are extremely sensitive to departures from normality.) Use histograms and normal quantile plots to verify normality.

STATDISK Procedure for Hypothesis Tests About Two Standard Deviations or Two Variances

1. For each of the two samples, obtain the sample size n and the sample standard deviation s. (If you have two samples of raw data, you can find these statistics using the Descriptive Statistics module. Also, if the available information includes sample sizes and sample *variances*, be sure to take the square root of the

sample variances to obtain the sample *standard deviations*.)

2. Select **Analysis** from the main menu.

3. Select **Hypothesis Testing** from the subdirectory.

4. Select **St Dev - Two Samples.**

5. Make these entries in the dialog box:

   -In the "claim" box, select the case corresponding to the claim being tested.

   -Enter a significance level, such as 0.05 or 0.01.

   -In the appropriate boxes, enter the sample size and sample standard deviation for the first sample, then do the same for the second sample.

6. Click on the **Evaluate** button to obtain the test results.

Section 9-5 in the textbook includes an example based on the following sample data.

	Regular Coke	**Regular Pepsi**
n	36	36
$\bar{x}$	0.81682	0.82410
s	0.007507	0.005701

The textbook describes the procedure for using a 0.05 significance level in testing the claim that the two populations of regular Coke and regular Pepsi have the same standard deviation. Using the above STATDISK procedure, the dialog box will be as follows. The textbook procedure requires that the sample with the larger variance be designated as Sample 1, but this is not necessary with STATDISK. STATDISK automatically does the required calculations and it correctly handles cases in which the first sample has a variance smaller than the second sample.

The STATDISK display shows a *P*–value of 0.1082. Because that *P*–value is greater than the significance level of 0.05, we fail to reject the null hypothesis. There does not appear to be a significant difference between the two population standard deviations. There is not sufficient evidence to support the claim that the populations have different standard deviations.

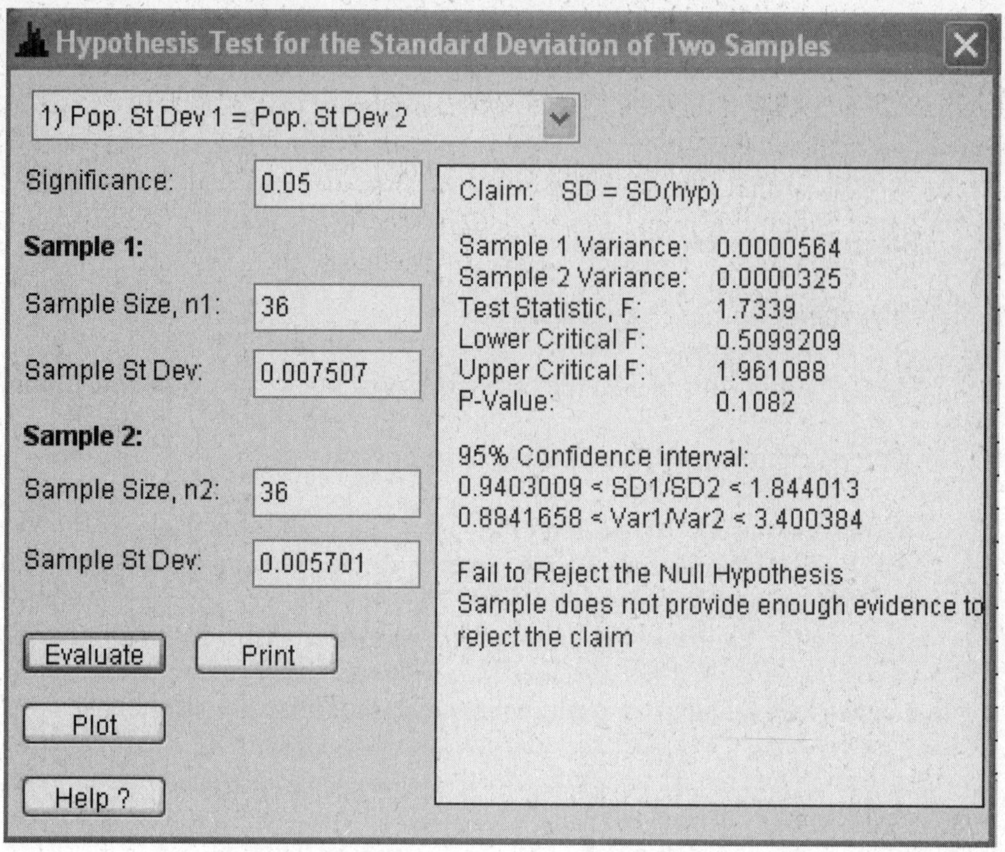

Hypothesis Test for the Standard Deviation of Two Samples

1) Pop. St Dev 1 = Pop. St Dev 2

Significance: 0.05

Sample 1:

Sample Size, n1: 36

Sample St Dev: 0.007507

Sample 2:

Sample Size, n2: 36

Sample St Dev: 0.005701

Evaluate | Print | Plot | Help ?

Claim: SD = SD(hyp)

Sample 1 Variance: 0.0000564
Sample 2 Variance: 0.0000325
Test Statistic, F: 1.7339
Lower Critical F: 0.5099209
Upper Critical F: 1.961088
P-Value: 0.1082

95% Confidence interval:
0.9403009 < SD1/SD2 < 1.844013
0.8841658 < Var1/Var2 < 3.400384

Fail to Reject the Null Hypothesis
Sample does not provide enough evidence to reject the claim

STATDISK Procedure for Confidence Interval Estimates of the Ratio of Two Standard Deviations or Variances

1. For each of the two samples, obtain the sample size *n*, and the sample standard deviation *s*. (If you have two samples of raw data, you can find these statistics using the Descriptive Statistics module. Also, if the available information includes sample sizes and sample *variances*, be sure to take the square root of the sample variances to obtain the sample *standard deviations*.)

2. Select **Analysis** from the main menu.

3. Select **Confidence Intervals** from the subdirectory.

4. Select **St Dev - Two Samples.**

5. Enter a confidence level and the sample statistics.

6. Click on the **Evaluate** button to obtain the test results.

CHAPTER 9 EXPERIMENTS: Inferences from Two Samples

9–1. ***E–Mail and Privacy*** A survey of 436 workers showed that 192 of them said that it was seriously unethical to monitor employee e–mail. When 121 senior–level bosses were surveyed, 40 said that it was seriously unethical to monitor employee e–mail (based on data from a Gallup poll). Use a 0.05 significance level to test the claim that for those saying that monitoring e–mail is seriously unethical, the proportion of employees is greater than the proportion of bosses.

Test statistic: __________ Critical value(s): __________ *P*–value: __________

Conclusion in your own words: ____

____

9–2. ***E–Mail and Privacy*** Refer to the sample data given in Experiment 9–1 and construct a 90% confidence interval estimate of the difference between the two population proportions. Is there a substantial gap between the employees and bosses?

____

____

9–3. ***Exercise and Coronary Heart Disease*** In a study of women and coronary heart disease, the following sample results were obtained: Among 10,239 women with a low level of physical activity (less than 200 kcal/wk), there were 101 cases of coronary heart disease. Among 9877 women with physical activity measured between 200 and 600 kcal/wk, there were 56 cases of coronary heart disease (based on data from "Physical Activity and Coronary Heart Disease in Women" by Lee, Rexrode, et al, *Journal of the American Medical Association,* Vol. 285, No. 11). Construct a 90% confidence interval estimate for the difference between the two proportions. Does the difference appear to be substantial? Does it appear that physical activity corresponds to a lower rate of coronary heart disease?

____

____

9–4. ***Exercise and Coronary Heart Disease*** Refer to the sample data in Experiment 9–3 and use a 0.05 significance level to test the claim that the rate of coronary heart disease is higher for women with the lower levels of physical activity.

Test statistic: __________ Critical value(s): __________ *P*–value: __________

(*continued*)

Conclusion in your own words: ______________________________

____

9–5. ***Instant Replay in Football*** In the 2000 football season, 247 plays were reviewed by officials using instant video replays, and 83 of them resulted in reversal of the original call. In the 2001 football season, 258 plays were reviewed and 89 of them were reversed (based on data from "Referees Turn to Video Aid More Often" by Richard Sandomir, *New York Times*). Is there a significant difference in the two reversal rates?

Test statistic: __________ Critical value(s): __________ *P*–value: __________

Conclusion in your own words: ______________________________

____

Does it appear that the reversal rate was the same in both years? __________

9–6. ***Effectiveness of Smoking Bans*** The Joint Commission on Accreditation of Healthcare Organizations mandated that hospitals ban smoking by 1994. In a study of the effects of this ban, subjects who smoke were randomly selected from two different populations. Among 843 smoking employees of hospitals with the smoking ban, 56 quit smoking one year after the ban. Among 703 smoking employees from workplaces without a smoking ban, 27 quit smoking a year after the ban (based on data from "Hospital Smoking Bans and Employee Smoking Behavior" by Longo, Brownson, et al, *Journal of the American Medical Association*, Vol. 275, No. 16). Is there a significant difference between the two proportions at a 0.05 significance level?

Test statistic: __________ Critical value(s): __________ *P*–value: __________

Conclusion in your own words: ______________________________

____

Is there a significant difference between the two proportions at a 0.01 significance level?

Test statistic: __________ Critical value(s): __________ *P*–value: __________

Conclusion in your own words: ______________________________

____

Does it appear that the ban had an effect on the smoking quit rate? __________

9–7. ***Testing Effectiveness of Vaccine*** In a *USA Today* article about an experimental nasal spray vaccine for children, the following statement was presented: "In a trial involving 1602 children only 14 (1%) of the 1070 who received the vaccine developed the flu, compared with 95 (18%) of the 532 who got a placebo." The article also referred to a study claiming that the experimental nasal spray "cuts children's chances of getting the flu." Is there sufficient sample evidence to support the stated claim?

Test statistic: __________ Critical value(s): __________ *P*–value: __________

Conclusion in your own words: ____

____

In Experiments 9–8 through 9–13, assume that the two samples are independent simple random samples selected from normally distributed populations. Do not assume that the population standard deviations are equal.

9–8. ***Hypothesis Test for Effect of Marijuana Use on College Students*** Many studies have been conducted to test the effects of marijuana use on mental abilities. In one such study, groups of light and heavy users of marijuana in college were tested for memory recall, with the results given below (based on data from "The Residual Cognitive Effects of Heavy Marijuana Use in College Students" by Pope and Yurgelun–Todd, *Journal of the American Medical Association*, Vol. 275, No. 7). Use a 0.01 significance level to test the claim that the population of heavy marijuana users has a lower mean than the light users.

Items sorted correctly by light marijuana users: $n = 64$, $\bar{x} = 53.3$, $s = 3.6$
Items sorted correctly by heavy marijuana users: $n = 65$, $\bar{x} = 51.3$, $s = 4.5$

Test statistic: __________ Critical value(s): __________ *P*–value: __________

Conclusion in your own words: ____

____

Based on these results, should marijuana use be of concern to college students?

____

9–9. ***Confidence Interval for Effects of Marijuana Use on College Students*** Refer to the sample data used in Experiment 9–8 and construct a 98% confidence interval for the difference between the two population means. Does the confidence interval include zero? What does the confidence interval suggest about the equality of the two population means?

____

9–10. ***Confidence Interval for Bipolar Depression Treatment*** In clinical experiments involving different groups of independent samples, it is important that the groups are similar in the important ways that affect the experiment. In an experiment designed to test the effectiveness of paroxetine for treating bipolar depression, subjects were measured using the Hamilton depression scale with the results given below (based on data from "Double–Blind, Placebo–Controlled Comparison of Imipramine and Paroxetine in the Treatment of Bipolar Depression" by Nemeroff, Evans, et al, *American Journal of Psychiatry*, Vol. 158, No. 6). Construct a 95% confidence interval for the difference between the two population means.

Placebo group: $n = 43$, $\bar{x} = 21.57$, $s = 3.87$
Paroxetine treatment group: $n = 33$, $\bar{x} = 20.38$, $s = 3.91$

____

Based on the results, does it appear that the two populations have different means? Should paroxetine be recommended as a treatment for bipolar depression?

____

____

9–11. ***Hypothesis Test for Bipolar Depression Treatment*** Refer to the sample data in Experiment 9–10 and use a 0.05 significance level to test the claim that the treatment group and placebo group come from populations with the same mean.

Test statistic: __________ Critical value(s): __________ *P*–value: __________

Conclusion in your own words: ______________________________

____

What does the result of the hypothesis test suggest about paroxetine as a treatment for bipolar depression?

____

____

9–12. ***Hypothesis Test for Magnet Treatment of Pain*** People spend huge sums of money (currently around $5 billion annually) for the purchase of magnets used to treat a wide variety of pains. Researchers conducted a study to determine whether magnets are effective in treating back pain. Pain was measured using the visual analog scale, and the results given below are among the results obtained in the study (based on data from "Bipolar Permanent Magnets for the Treatment of Chronic Lower Back Pain: A Pilot Study" by Collacott, Zimmerman, White, and Rindone, *Journal of the American Medical*

Association, Vol. 283, No. 10). Use a 0.05 significance level to test the claim that those treated with magnets have a greater reduction in pain than those given a sham treatment (similar to a placebo).

Reduction in pain level after magnet treatment: $n = 20$, $\bar{x} = 0.49$, $s = 0.96$
Reduction in pain level after sham treatment: $n = 20$, $\bar{x} = 0.44$, $s = 1.4$

Test statistic: __________ Critical value(s): __________ *P*–value: __________

Conclusion in your own words: ____

____

Does it appear that magnets are effective in treating back pain? Is it valid to argue that magnets might appear to be effective if the sample sizes are larger?

____

____

9–13. ***Confidence Interval for Magnet Treatment of Pain*** Refer to the sample data from Experiment 9–12 and construct a 90% confidence interval estimate of the difference between the mean reduction in pain for those treated with magnets and the mean reduction in pain for those given a sham treatment. Based on the result, does it appear that the magnets are effective in reducing pain?

____

9–14. ***Self–Reported and Measured Female Heights*** As part of the National Health and Nutrition Examination Survey conducted by the Department of Health and Human Services, self–reported heights and measured heights were obtained for females aged 12–16. Listed below are sample results.

Reported height	53	64	61	66	64	65	68	63	64	64	64	67
Measured height	58.1	62.7	61.1	64.8	63.2	66.4	67.6	63.5	66.8	63.9	62.1	68.5

Is there sufficient evidence to support the claim that there is a difference between self–reported heights and measured heights of females aged 12–16? Use a 0.05 significance level.

Test statistic: __________ Critical value(s): __________ *P*–value: __________

(continued)

Conclusion in your own words: ______________________________

____

Construct a 95% confidence interval estimate of the mean difference between reported heights and measured heights. Interpret the resulting confidence interval, and comment on the implications of whether the confidence interval limits contain 0.

____

____

9–15. ***Self–Reported and Measured Male Heights*** As part of the National Health and Nutrition Examination Survey conducted by the Department of Health and Human Services, self–reported heights and measured heights were obtained for males aged 12–16. Listed below are sample results.

Reported height	68	71	63	70	71	60	65	64	54	63	66	72
Measured height	67.9	69.9	64.9	68.3	70.3	60.6	64.5	67.0	55.6	74.2	65.0	70.8

Is there sufficient evidence to support the claim that there is a difference between self–reported heights and measured heights of males aged 12–16? Use a 0.05 significance level.

Test statistic: __________ Critical value(s): __________ *P*–value: __________

Conclusion in your own words: ______________________________

____

Construct a 95% confidence interval estimate of the mean difference between reported heights and measured heights. Interpret the resulting confidence interval, and comment on the implications of whether the confidence interval limits contain 0.

____

____

9–16. ***Effectiveness of SAT Course*** Refer to the data in the table that lists SAT scores before and after the sample of 10 students took a preparatory course (based on data from the College Board and "An Analysis of the Impact of Commercial Test Preparation Courses on SAT Scores," by Sesnowitz, Bernhardt, and Knain, *American Educational Research Journal,* Vol. 19, No. 3.)

(*continued*)

Student	A	B	C	D	E	F	G	H	I	J
SAT score before course (x)	700	840	830	860	840	690	830	1180	930	1070
SAT score after course (y)	720	840	820	900	870	700	800	1200	950	1080

Is there sufficient evidence to conclude that the preparatory course is effective in raising scores? Use a 0.05 significance level.

Test statistic: __________ Critical value(s): __________ *P*–value: __________

Conclusion in your own words: ______________________________

____

Construct a 95% confidence interval estimate of the mean difference between the before and after scores. Write a statement that interprets the resulting confidence interval.

____

9–17. ***Before/After Treatment Results*** Captopril is a drug designed to lower systolic blood pressure. When subjects were tested with this drug, their systolic blood pressure readings (in mm of mercury) were measured before and after the drug was taken, with the results given in the accompanying table (based on data from "Essential Hypertension: Effect of an Oral Inhibitor of Angiotensin-Converting Enzyme," by MacGregor et al., *British Medical Journal,* Vol. 2).

Subject	A	B	C	D	E	F	G	H	I	J	K	L
Before	200	174	198	170	179	182	193	209	185	155	169	210
After	191	170	177	167	159	151	176	183	159	145	146	177

Use the sample data to construct a 99% confidence interval for the mean difference between the before and after readings.

____

Is there sufficient evidence to support the claim that captopril is effective in lowering systolic blood pressure?

____

9–18. ***Ages of Faculty and Student Cars*** Students at the author's college randomly selected 217 student cars and found that they had ages with a mean of 7.89 years and a standard deviation of 3.67 years. They also randomly selected 152 faculty cars and found that they had ages with a mean of 5.99 years and a standard deviation of 3.65 years. Is there sufficient evidence to support the claim that the ages of faculty cars vary less than the ages of student cars?

Test statistic: __________ Critical value(s): __________ *P*–value: __________

Conclusion in your own words: ____

____

9–19. ***Testing Effects of Zinc*** A study of zinc-deficient mothers was conducted to determine effects of zinc supplementation during pregnancy. Sample data are listed below (based on data from "The Effect of Zinc Supplementation on Pregnancy Outcome," by Goldenberg et al., *Journal of the American Medical Association*, Vol. 274, No. 6). The weights were measured in grams. Using a 0.05 significance level, is there sufficient evidence to support the claim that the variation of birth weights for the placebo population is greater than the variation for the population treated with zinc supplements?

Zinc Supplement Group	Placebo Group
$n = 294$	$n = 286$
$\bar{x} = 3214$	$\bar{x} = 3088$
$s = 669$	$s = 728$

Test statistic: __________ Critical value(s): __________ *P*–value: __________

Conclusion in your own words: ____

____

9–20. ***Rainfall on Weekends*** *USA Today* and other newspapers reported on a study that supposedly showed that it rains more on weekends. The study referred to areas on the east coast near the ocean. Data Set 10 in Appendix B lists the rainfall amounts in Boston for one year. The first 52 rainfall amounts for Wednesday have a mean of 0.0517 in. and a standard deviation of 0.1357 in. The 52 rainfall amounts for Sunday have a mean of 0.0677 in. and a standard deviation of 0.2000 in.

Assuming that we want to use the methods of Section 9–5 in the textbook to test the claim that Wednesday and Sunday rainfall amounts have the same standard deviation, identify the *F* test statistic, critical value, and conclusion. Use a 0.05 significance level. Enter the results on the following page.

Test statistic: __________ Critical value(s): __________ *P*–value: __________

Conclusion in your own words: ____

____

Consider the prerequisite of normally distributed populations. Instead of constructing histograms or normal quantile plots, simply examine the numbers of days with no rainfall. Are Wednesday rainfall amounts normally distributed? Are Sunday rainfall amounts normally distributed? What can be concluded from these results?

____

____

9–21. ***Tobacco and Alcohol Use in Animated Children's Movies*** Data Set 5 in Appendix B of the textbook lists times (in seconds) that animated children's movies show tobacco use and alcohol use. The 50 times of tobacco use have a mean of 57.4 sec and a standard deviation of 104.0 sec. The 50 times of alcohol use have a mean of 32.46 sec and a standard deviation of 66.3 sec. Assuming that we want to use the methods of this section to test the claim that the times of tobacco use and the times of alcohol use have different standard deviations, identify the F test statistic, critical value, and conclusion. Use a 0.05 significance level.

Test statistic: __________ Critical value(s): __________ *P*–value: __________

Conclusion in your own words: ____

____

Consider the prerequisite of normally distributed populations. Instead of constructing histograms or normal quantile plots, simply examine the numbers of movies showing no tobacco or alcohol use. Are the times for tobacco use normally distributed? Are the times for alcohol use normally distributed? What can be concluded from these results?

____

____

9–22. ***Calcium and Blood Pressure*** Sample data were collected in a study of calcium supplements and their effects on blood pressure. A placebo group and a calcium group began the study with blood pressure measurements (based on data from “Blood Pressure and Metabolic Effects of Calcium Supplementation in Normotensive White and Black Men,” by Lyle et al., *Journal of the American Medical Association,* Vol. 257, No. 13). Sample values are listed below. At the 0.05 significance level, test the claim that the two sample groups come from populations with the same standard deviation.

Placebo:	124.6	104.8	96.5	116.3	106.1	128.8	107.2	123.1
	118.1	108.5	120.4	122.5	113.6			
Calcium:	129.1	123.4	102.7	118.1	114.7	120.9	104.4	116.3
	109.6	127.7	108.0	124.3	106.6	121.4	113.2	

Test statistic: __________ Critical value(s): __________ *P*–value: __________

Conclusion in your own words: ______________________________

____

If the experiment requires groups with equal standard deviations, are these two groups acceptable?

9–23. ***Weights of Quarters*** Weights of quarters are used by vending machines as one way to detect counterfeit coins. Data Set 14 in Appendix B includes weights of pre-1964 silver quarters and post 1964 quarters. Here are the summary statistics: Pre-1964: $n = 40$, $\bar{x} = 6.19267$ g, $s = 0.08700$ g; post-1964: $n = 40$, $\bar{x} = 5.63930$ g, $s = 0.06194$. Use a 0.05 significance level to test the claim that the two populations of quarters have the same standard deviation. If the amounts of variation are different, vending machines might need more complicated adjustments. Does it appear that such adjustments are necessary?

Test statistic: __________ Critical value(s): __________ *P*–value: __________

Conclusion in your own words: ______________________________

____

9–24. ***Weights of Pennies and Quarters*** Data Set 14 in Appendix B includes weights of post-1983 pennies and post-1964 quarters. Here are the summary statistics: Post-1983 pennies: $n = 37$, $\bar{x} = 2.49910$ g, $s = 0.01648$ g; post 1964 quarters: $n = 40$, $\bar{x} = 5.63930$ g, $s = 0.06194$. Test the claim that post-1983 pennies and post-1964 quarters have the same amount of variation. Should they have the same amount of variation?

Test statistic: __________ Critical value(s): __________ *P*–value: __________

Conclusion in your own words: ______________________________

____

9–25. ***Ages of Stowaways*** The ages of stowaways on the Queen Mary, categorized by westbound crossings and eastbound crossings, are given below. The data are from the Cunard Steamship Co., Ltd. Test the claim that the ages are from populations having the same mean.

Test statistic: __________ Significance level: __________

Critical value(s): __________ *P*–value: __________

Conclusion in your own words: ____

____

Westbound:

41 24 32 26 39 45 24 21 22 21 40 18 33 33 19 31 16 16 23 19 16 20 18
22 26 22 38 42 25 21 29 24 18 17 24 18 19 30 18 24 31 30 48 29 34 25
23 41 16 17 15 19 18 66 27 43

Eastbound:

24 24 34 15 19 22 18 20 20 17 17 20 18 23 37 15 25 28 21 15 48 18 12
15 23 25 22 21 30 19 20 20 35 19 38 26 19 20 19 41 31 20 19 18 42 25
19 47 19 22 20 23 24 37 23 30 32 28 32 48 27 31 22 34 26 20 22 15 19
20 18 26 36 31 35

Correlation and Regression

10-1 Linear Correlation and Regression

Sections 10-2, 10-3, and 10-4 in the textbook introduce the basic concepts of linear correlation and regression. The basic objective is to use sample paired data to determine whether there is a relationship between two variables and, if so, identify what the relationship is. Consider the paired data in the table below. We want to determine whether there is a correlation between the duration time of an eruption and the time interval after that eruption. The STATDISK procedure for investigation correlation follows.

Eruptions of the Old Faithful Geyser

Duration	240	120	178	234	235	269	255	220
Interval After	92	65	72	94	83	94	101	87

STATDISK Procedure for Correlation and Regression

1. Either enter the data in columns of the Statdisk Data Window, or open files that already exist.
2. Select **Analysis** from the main menu.
3. Select the menu item of **Correlation and Regression**.
4. Select a significance level, such as 0.05 or 0.01.
5. Proceed to select the columns to be used for the x variable and the y variable.
6. Click the **Evaluate** button to get the correlation/regression results.
7. Click the **Scatterplot** button to get a graph of the scatterplot.

Correlation If you follow the above steps using the sample data in the above table, the STATDISK results will be as shown on the following page. Also shown is the scatterplot that is obtained by clicking on the **Plot** button in the display window. The results include the linear correlation coefficient of $r = 0.926$ (rounded), the critical values of $r = \pm 0.707$, the P-value of 0.00097, and the conclusion that there is sufficient evidence to reject the null hypothesis (of no correlation) and support a claim of a linear correlation between the two variables. There does appear to be a linear correlation between the duration time and the time interval after an eruption.

Regression Also included in the display on the next page are the y-intercept b_0 and slope b_1 of the estimated regression line. Using the STATDISK results, the estimated regression equation is

$$\hat{y} = 34.8 + 0.234x$$

The graph of the scatter diagram (shown on the next page) includes the regression line.

Correlation and Regression

Significance: 0.05

Select the columns to be used for the x and y variables.

x variable column: 1   y variable column: 2

Sample size, n:	8
Degrees of freedom:	6
Correlation Results:	
Correlation coeff, r:	0.9255912
Critical r:	±0.706734
P-value (two-tailed):	0.00097

Reject the Null Hypothesis
Sample provides evidence to support linear correlation

Regression Results:
Y= b0 + b1x:

Y Intercept, b0:	34.7698
Slope, b1:	0.2340614
Total Variation:	1036
Explained Variation:	887.5609
Unexplained Variation:	148.4391
Standard Error:	4.973916
Coeff of Det, R^2:	0.8567191

Evaluate   Print   Plot   Help ?

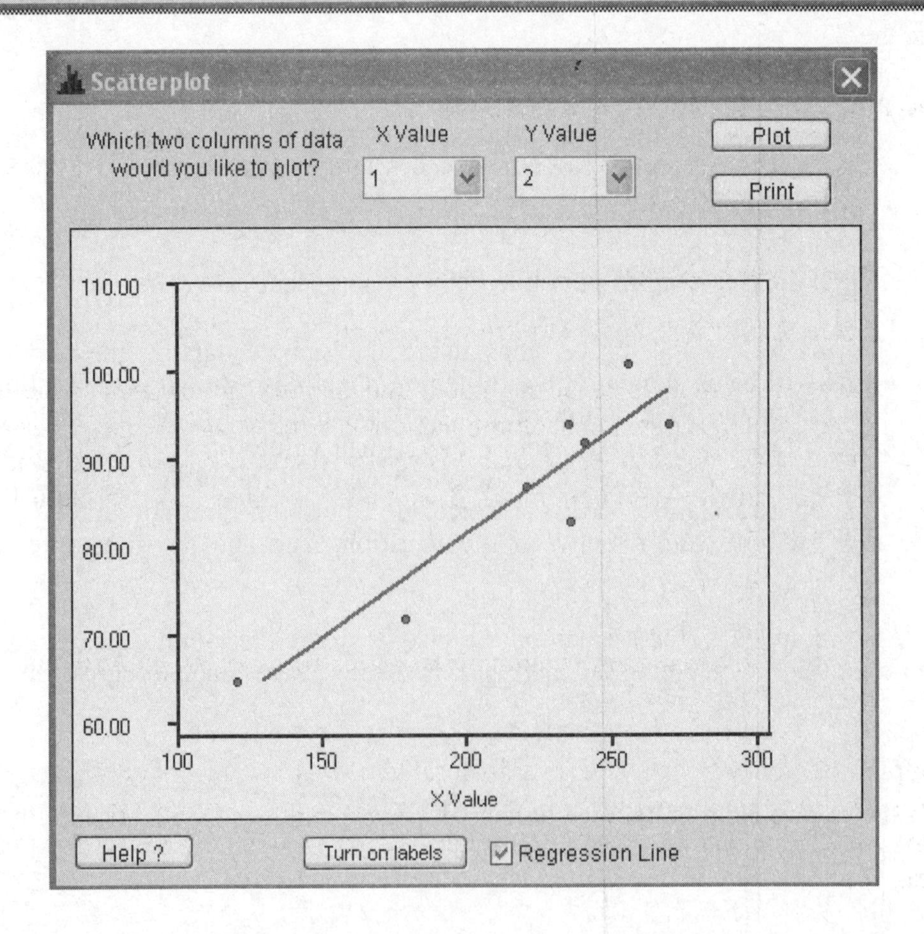

In the preceding display of the scatterplot, note that in addition to the Help button, there are two other buttons. Their functions are described as follows.

Turn labels on: Clicking this button causes the coordinates of the points to be shown in the graph. Each pair of coordinates is positioned near the point it represents.

Regression line: The default is indicated by the check mark in the box for the "Regression line" button. If you click on that box to remove the check mark, the graph of the regression line will not be included in the scatterplot.

10-2 Multiple Regression

Section 10-5 of *Elementary Statistics* discusses multiple regression, and STATDISK does allow you to obtain multiple regression results. Once a collection of sample data has been entered, you can easily experiment with different combinations of columns to find the combination that is best. Here is the STATDISK procedure.

STATDISK Procedure for Multiple Regression

1. Either enter the data in columns of the Statdisk Data Window, or open a file that already exists.
2. Select **Analysis** from the main menu.
3. Select **Multiple Regression** from the menu.
4. Select the columns to be included, and identify the column to be used for the dependent variable. (If a column has a check mark but you don't want it included, click on the check mark so that it is removed and its column is excluded.) See the screen below, where columns 1, 3, and 6 are included, with column 1 selected for the dependent variable.
5. Click on **Evaluate**.
6. To use a different combination of variables, simply click on different combinations of columns.

As an example, see the table of Old Faithful eruption data on the following page. If we use STATDISK's **Multiple Regression** module with "interval after" time as the response *y* variable and predictor variables of duration time and height, we get the STATDISK results that follow the table of data.

Eruptions of the Old Faithful Geyser

Duration (x_1)	240	120	178	234	235	269	255	220
Height (x_2)	140	110	125	120	140	120	125	150
Interval After (y)	92	65	72	94	83	94	101	87

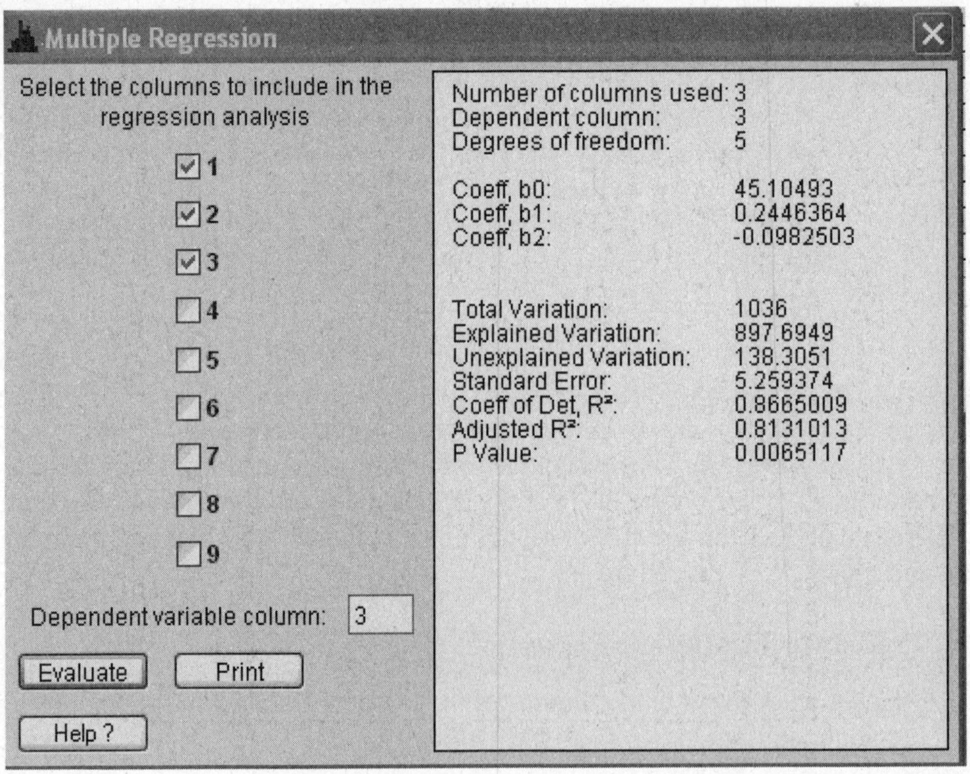

Multiple Regression

Select the columns to include in the regression analysis

☑ 1
☑ 2
☑ 3
☐ 4
☐ 5
☐ 6
☐ 7
☐ 8
☐ 9

Dependent variable column: 3

Evaluate Print

Help ?

Number of columns used:	3
Dependent column:	3
Degrees of freedom:	5
Coeff, b0:	45.10493
Coeff, b1:	0.2446364
Coeff, b2:	-0.0982503
Total Variation:	1036
Explained Variation:	897.6949
Unexplained Variation:	138.3051
Standard Error:	5.259374
Coeff of Det, R²:	0.8665009
Adjusted R²:	0.8131013
P Value:	0.0065117

The results in the above STATDISK display include the intercept $b_0 = 45.1$ (rounded), the coefficient b1 = 0.245 (rounded), and the coefficient b2 = −0.0983 (rounded). These values are included in the multiple regression equation as shown here:

$$\hat{y} = 45.1 + 0.245x_1 - 0.0983x_2$$

The results also include the adjusted coefficient of determination (Adjusted $R^2 = 0.813$), as well as the P-value of 0.0065117. The small P-value suggests that the multiple regression equation is a good model for the data.

10-3 Modeling

Section 10-6 in *Elementary Statistics* discusses mathematical modeling. The objective is to find a mathematical function that "fits" or describes real-world data. Among the models discussed in the textbook, we will describe how STATDISK can be used for the linear, quadratic, logarithmic, exponential, and power models.

Consider the sample data in Table 10-4 from the textbook, reproduced below. As in the textbook, we use the coded year values for x, so $x = 1, 2, 3, \ldots, 11$. The y values are the populations (in millions) of 5, 10, 17, . . . , 281. The STATDISK scatter diagram is displayed.

TABLE 10-4 **Population (in millions) of the United States**

Year	1800	1820	1840	1860	1880	1900	1920	1940	1960	1980	2000
Coded year	1	2	3	4	5	6	7	8	9	10	11
Population	5	10	17	31	50	76	106	132	179	227	281

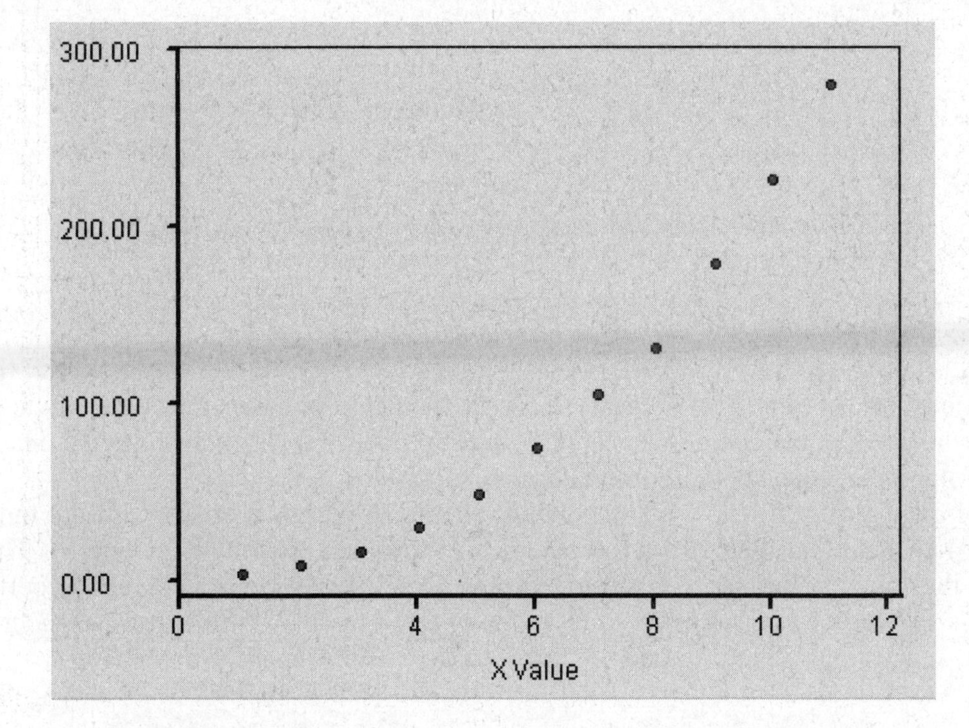

Linear Model: $y = a + bx$

The linear model can be obtained by using STATDISK's correlation and regression module. The procedure is described in Section 10-1 of this manual/workbook. Select **Analysis**, then **Correlation and Regression**. For the data in Table 10-4, enter the coded year values of 1, 2, 3, . . . , 11 for the first column and enter the population values of 5, 10, 17, . . . , 281 for the second column. The result will be as shown below.

The following STATDISK display describes key results for the linear model. The resulting function is $y = -61.92727 + 27.2x$. The coefficient of determination is displayed as $r^2 = 0.9245459$. The high value of r^2 suggests that the linear model is a reasonably good fit.

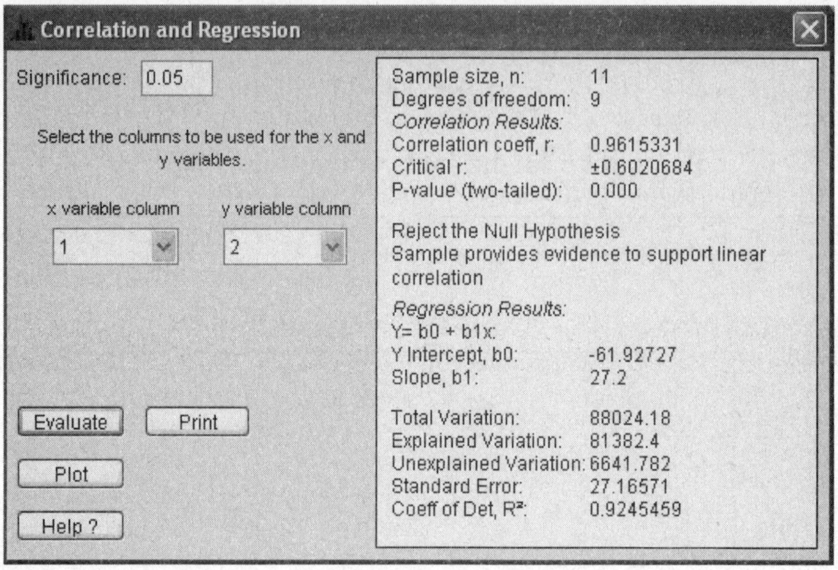

Correlation and Regression

Significance: 0.05

Select the columns to be used for the x and y variables.

x variable column: 1 | y variable column: 2

Evaluate | Print | Plot | Help ?

Sample size, n: 11
Degrees of freedom: 9
Correlation Results:
Correlation coeff, r: 0.9615331
Critical r: ±0.6020684
P-value (two-tailed): 0.000

Reject the Null Hypothesis
Sample provides evidence to support linear correlation

Regression Results:
Y= b0 + b1x:
Y Intercept, b0: -61.92727
Slope, b1: 27.2

Total Variation: 88024.18
Explained Variation: 81382.4
Unexplained Variation: 6641.782
Standard Error: 27.16571
Coeff of Det, R²: 0.9245459

Quadratic Model: $y = ax^2 + bx + c$

In the Statdisk Data Window, enter the values of x and y in the first two columns, and the values of x^2 in the third column. Select **Analysis**, then **Multiple Regression.** When indicating the columns to be used, select columns 1, 2, and 3, and be sure to identify the column containing the y values (column 2) as the column for the dependent variable.

The display shown below corresponds to the quadratic model used with the sample data in Table 10-4. Note that there are three columns of data representing x, y, and x^2. The results show that the function has the form given as $y = 10.0 - 0.600x + 2.77x^2$. (The coefficient b_1 corresponds to x and b_2 corresponds to x^2.) The coefficient of determination is given by $R^2 = 0.9991688$, suggesting a better fit than the linear model (which has $R^2 = 0.9245459$).

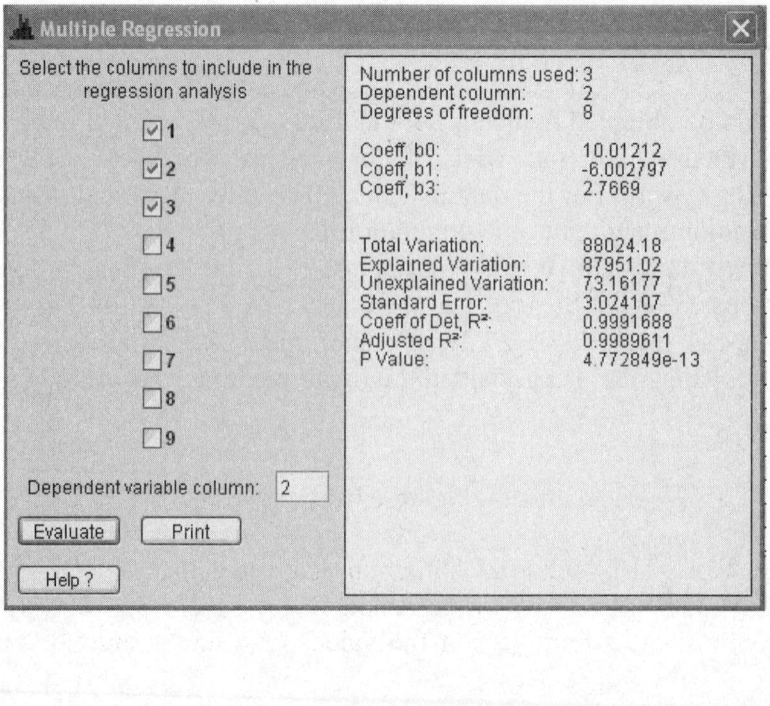

Multiple Regression

Select the columns to include in the regression analysis

☑1 ☑2 ☑3 ☐4 ☐5 ☐6 ☐7 ☐8 ☐9

Dependent variable column: 2

Evaluate | Print | Help ?

Number of columns used: 3
Dependent column: 2
Degrees of freedom: 8

Coeff, b0: 10.01212
Coeff, b1: -6.002797
Coeff, b3: 2.7669

Total Variation: 88024.18
Explained Variation: 87951.02
Unexplained Variation: 73.16177
Standard Error: 3.024107
Coeff of Det, R²: 0.9991688
Adjusted R²: 0.9989611
P Value: 4.772849e-13

Logarithmic Model: $y = a + b \ln x$

In the Statdisk Data Window, enter the values of ln x in the first column and enter the values of y in the second column. (Values of ln x can be found using a calculator or STATDISK's **Sample Transformations** feature.) Select **Analysis**, then **Correlation and Regression**.

The display shown below results from the logarithmic model used with the sample data in Table 10-4. The function is given by $y = -65.9 + 105 \ln x$, with $R^2 = 0.6961498$, suggesting that this model does not fit as well as the linear or quadratic models. Of the three models considered so far, the quadratic model appears to be best (because it has the highest value of R^2).

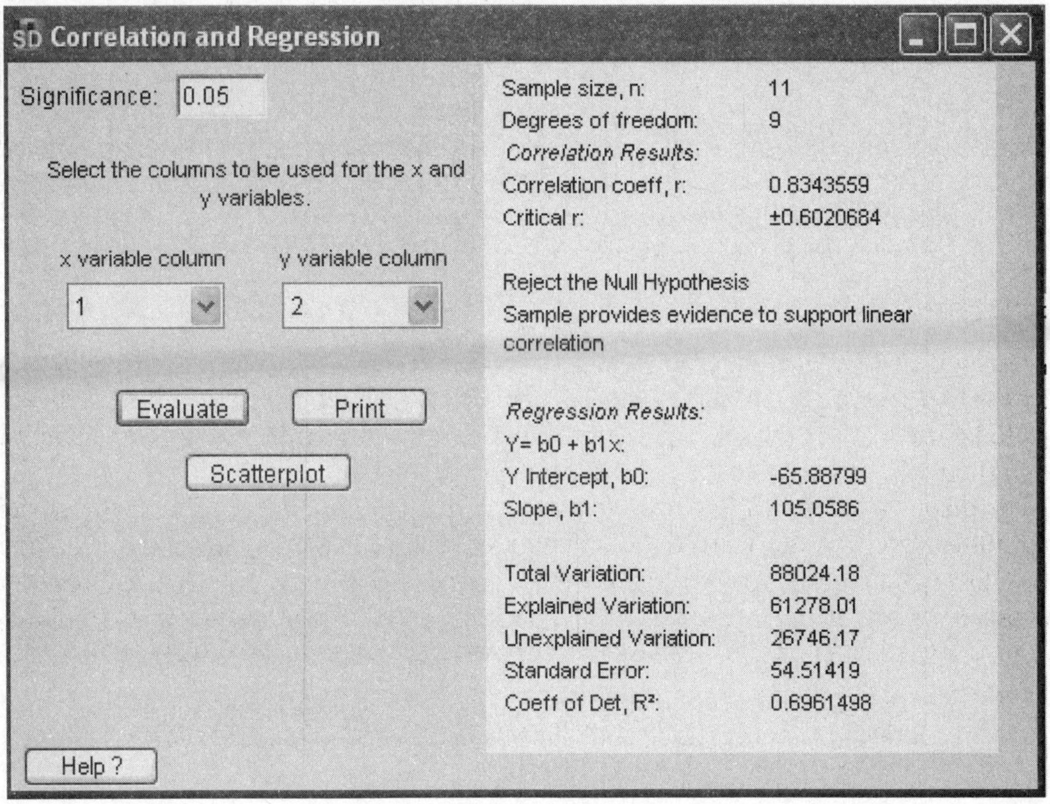

Correlation and Regression

Significance: 0.05

Select the columns to be used for the x and y variables.

x variable column: 1 | y variable column: 2

Evaluate | Print | Scatterplot | Help ?

Sample size, n:	11
Degrees of freedom:	9
Correlation Results:	
Correlation coeff, r:	0.8343559
Critical r:	±0.6020684

Reject the Null Hypothesis
Sample provides evidence to support linear correlation

Regression Results:	
Y= b0 + b1x:	
Y Intercept, b0:	-65.88799
Slope, b1:	105.0586
Total Variation:	88024.18
Explained Variation:	61278.01
Unexplained Variation:	26746.17
Standard Error:	54.51419
Coeff of Det, R²:	0.6961498

Exponential Model: $y = ab^x$

The exponential model is tricky, but it can be obtained using STATDISK. Enter the values of x in the first column, and enter the values of ln y in the second column. Select **Analysis**, then **Correlation and Regression**. When you get the results from STATDISK, the value of the coefficient of determination is correct, but the values of a and b in the exponential model must be computed as follows:

To find the value of a: Evaluate e^{b0} where b_0 is given by STATDISK.

To find the value of b: Evaluate e^{b1} where b_1 is given by STATDISK.

Using the data in Table 10-4, this procedure results in the STATDISK display that follows. The value of $R^2 = 0.9631106$ is OK as is, but the values of a and b must be computed from the

STATDISK results as shown below:

$$a = e^{b0} = e^{1.655597} = 5.2362$$

$$b = e^{b1} = e^{0.39440506} = 1.4830$$

Using these values of a and b, we express the exponential model as $y = 5.2362(1.4830^x)$.

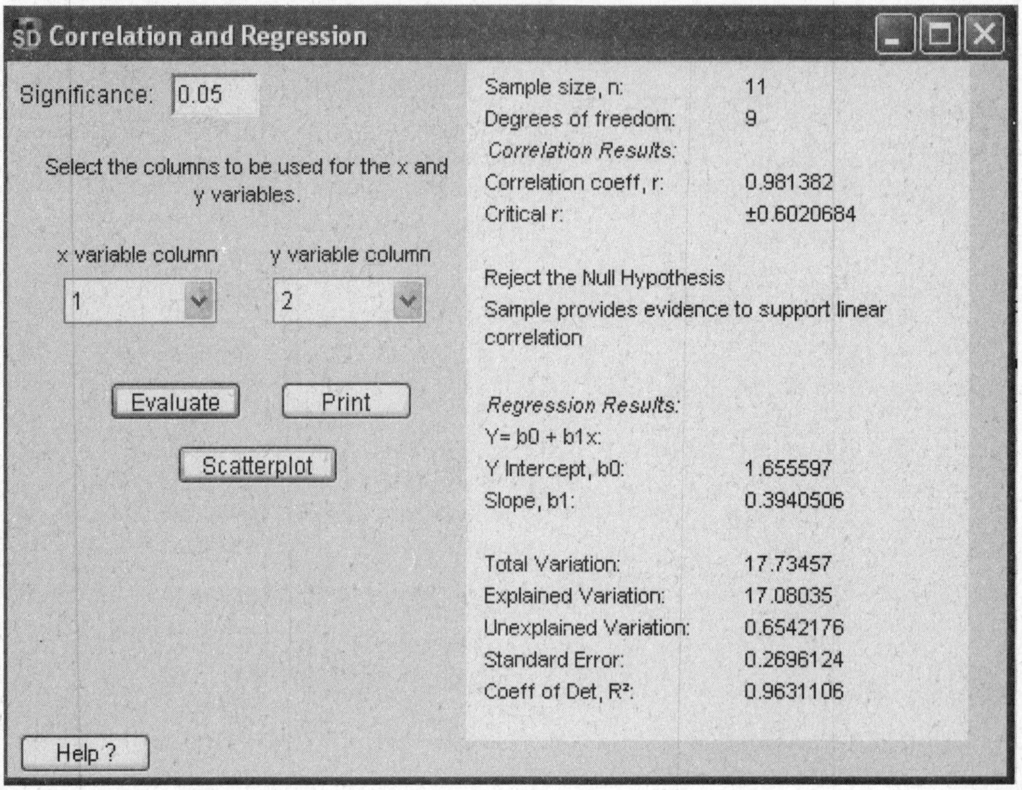

SD Correlation and Regression

Significance: 0.05

Select the columns to be used for the x and y variables.

x variable column: 1
y variable column: 2

Evaluate | Print | Scatterplot

Sample size, n:	11
Degrees of freedom:	9
Correlation Results:	
Correlation coeff, r:	0.981382
Critical r:	±0.6020684

Reject the Null Hypothesis
Sample provides evidence to support linear correlation

Regression Results:	
Y= b0 + b1x:	
Y Intercept, b0:	1.655597
Slope, b1:	0.3940506
Total Variation:	17.73457
Explained Variation:	17.08035
Unexplained Variation:	0.6542176
Standard Error:	0.2696124
Coeff of Det, R²:	0.9631106

Help ?

Power Model: $y = ax^b$

The power model is also tricky, but it can be obtained using STATDISK. Enter the values of ln x in the first column, and enter the values of ln y in the second column. Select **Analysis**, then **Correlation and Regression**. When you get the results from STATDISK, the value of the coefficient of determination is correct, but the values of a and b in the power model are found as follows:

To find the value of a: Evaluate e^{b0} where b_0 is given by STATDISK.

The value of b is the same as the value of b_1 given by STATDISK.

Using the data in Table 10-4, this procedure results in the STATDISK display shown below. The value of $R^2 = 0.9764065$ is OK as is, and it suggests that the power model is not as good as the quadratic model. The values of *a* and *b* are found from the STATDISK results as shown below:

$a = e^{b0} = e^{1.209893} = 3.3531$
$b = b_1$ from STATDISK $= 1.7661$

Using these values of *a* and *b*, we express the power model as

$$y = 3.3531(x^{1.7661})$$

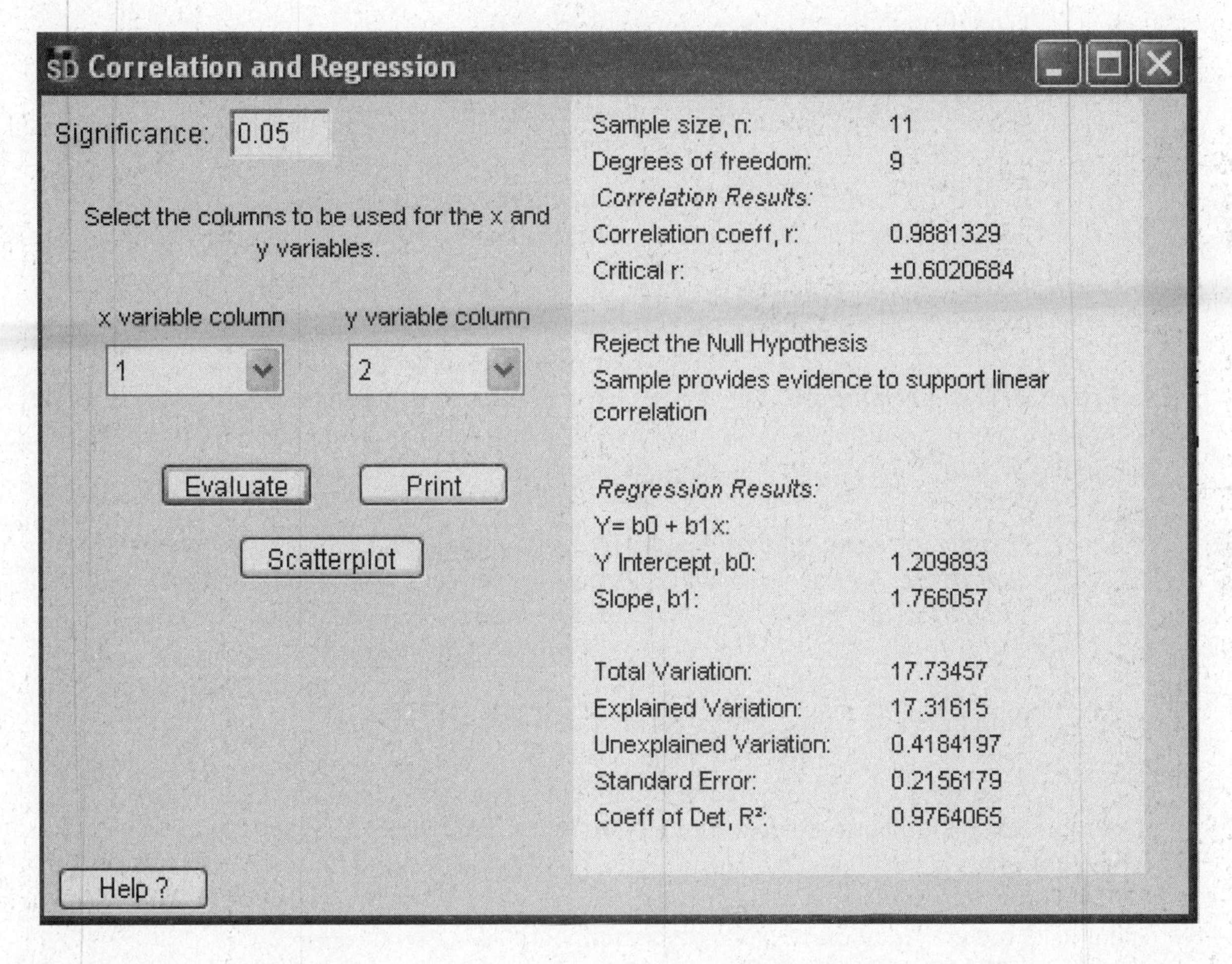

The rationale underlying the methods for the exponential and power models is based on transformations of equations. In the exponential model of $y = ab^x$, for example, taking natural logarithms of both sides yields $\ln y = \ln a + x (\ln b)$, which is the equation of a straight line. STATDISK can be used to find the equation of this straight line that fits the data best; the intercept will be $\ln a$ and the slope will be $\ln b$, but we need the values of *a* and *b*, so we solve for them as described above. Similar reasoning is used with the power model.

CHAPTER 10 EXPERIMENTS: Correlation and Regression

10–1. ***Bear Ages and Weights*** Open the BEARS data set and use the values for age (x) and the values for weight (y) to find the following. (Ages are in months and weights are in pounds.)

a. Display the scatterplot of the paired age/weight data. Based on that scatterplot, does there appear to be a relationship between the ages of bears and their weights? If so, what is it?

____

b. Find the value of the linear correlation coefficient r. ________

c. Assuming a 0.05 level of significance, what do you conclude about the correlation between ages and weights of bears?

____

d. Find the equation of the regression line. (Use age as the x predictor variable, and use weight as the y response variable.) _______________

e. What is the best predicted weight of a bear that is 36 months old?__________
(*Caution*: See the "Prediction" subsection in Section 10-3 of the textbook.)

10–2. ***Effect of Transforming Data*** The ages used in Experiment 10–1 are in months. Convert them to days by multiplying each age by 30, then repeat Experiment 10–1 and enter the responses here:

a. Display the scatterplot of the paired age/weight data. Based on that scatterplot, does there appear to be a relationship between the ages of bears and their weights? If so, what is it?

____

b. Find the value of the linear correlation coefficient r.______________

c. Assuming a 0.05 level of significance, what do you conclude about the correlation between ages and weights of bears?

____

d. Find the equation of the regression line. (Use age as the x predictor variable, and use weight as the y response variable.) _______________

e. What is the best predicted weight of a bear that is 36 months old?__________
(*Caution*: See the "Prediction" subsection in Section 10-3 of the textbook.)

f. After comparing the responses obtained in Experiment 10-1 to those obtained here, describe the general effect of changing the scale for one of the variables.

____

10–3. ***Bear Chest Sizes and Weights*** Open the BEARS data set and use the values for the chest sizes (x) and the values for weights (y) to find the following.

a. Display the scatter diagram of the paired chest size/weight data. Based on that scatter diagram, does there appear to be a relationship between the chest sizes of bears and their weights? If so, what is it?

____

b. Find the value of the linear correlation coefficient r. _________

c. Assuming a 0.05 level of significance, what do you conclude about the correlation between chest sizes and weights of bears?

____

d. Find the equation of the regression line. (Use chest size as the x predictor variable, and use weight as the y response variable.) ____________________

e. What is the best predicted weight of a bear that has a chest size of 40 in.?_____ (*Caution*: See the "Prediction" subsection in Section 10-3 of the textbook.)

10–4. ***Predicting Weight of a Bear*** Experiments 10-1 and 10-3 both used weight as the response variable. Experiment 10-1 used the predictor variable of age and Experiment 10-3 used the predictor variable of chest size. Open the BEARS data set and find the single variable that is best for predicting the weight of a bear. Identify that variable and give reasons for its selection.

____

____

____

10–5. ***Effect of No Variation for a Variable*** Use the following paired data and obtain the indicated results.

x	1	2	3	4	5	7	7	9
y	5	5	5	5	5	5	5	5

a. Print a scatterplot of the paired x and y data. Based on the result, does there appear to be a relationship between x and y? If so, what is it?

____

b. What happens when you try to find the value of r? Why?

____

c. What do you conclude about the correlation between x and y? What is the equation of the regression line?

____

10–6. ***Song Audiences and Sales*** The table below lists the numbers of audience impressions (in hundreds of millions) listening to songs and the corresponding numbers of albums sold (in hundreds of thousands). The number of audience impressions is a count of the number of times people have heard the song. The table is based on data from *USA Today*.

Audience impressions	28	13	14	24	20	18	14	24	17
Albums sold	19	7	7	20	6	4	5	25	12

Is there a correlation between audience impressions and albums sold? Explain.

____

What is the equation of the regression line? (Let the predictor variable represent audience impressions.)______________________________

Find the best predicted number of albums sold for a song with 20 (hundred million) audience impressions.______________________

10–7. ***Supermodel Heights and Weights*** Listed below are heights (in inches) and weights (in pounds) for supermodels Michelle Alves, Nadia Avermann, Paris Hilton, Kelly Dyer, Christy Turlington, Bridget Hall, Naomi Campbell, Valerie Mazza, and Kristy Hume.

Height	70	70.5	68	65	70	70	70	70	71
Weight	117	119	105	115	119	127	113	123	115

Is there a correlation between height and weight? If there is a correlation, does it mean that there is a correlation between height and weight of all adult women?

____

What is the equation of the regression line? ______________________

Find the best predicted weight of a supermodel who is 72 in. tall. ______________

10–8. ***Garbage Data for Predicting Household Size*** Data Set 16 in Appendix B of the textbook consists of data from the Garbage Project at the University of Arizona. Use household size as the response y variable. For each given predictor x variable, find the value of the linear correlation coefficient, the equation of the regression line, and the value of the coefficient of determination r^2. Enter the results on the following page.

	r	Equation of regression line	r^2
Metal	___	____________________	___
Paper	___	____________________	___
Plastic	___	____________________	___
Glass	___	____________________	___
Food	___	____________________	___
Yard	___	____________________	___
Text	___	____________________	___
Other	___	____________________	___
Total	___	____________________	___

Based on the above results, which single independent variable appears to be the best predictor of household size? Why?

____

10–9. ***Multiple Regression*** Use the same data set described in Experiment 10-8. Let household size be the dependent y variable and use the given predictor x variables to enter the results below.

	Multiple regression equation	R^2	Adj. R^2
Metal and Paper	____________________	___	____
Plastic and Food	____________________	___	____
Metal, Paper, Glass	____________________	___	____
Metal, Paper, Plastic, Glass	____________________	___	____

Based on the above results, which of the multiple regression equations appears to best fit the data? Why?

____

10–10. ***Cigarette Data*** Use Data Set 3 from Appendix B of the textbook. Assume that we want to predict the amount of nicotine in a cigarette, based on the amount of tar and carbon monoxide. Use nicotine as the dependent variable and use tar and/or CO (carbon monoxide) for independent variables. Find the equation that is best for predicting nicotine in a cigarette and describe why it is best.

____

____

10-11. ***Predicting Home Selling Price*** Refer to Data Set 18 in Appendix B in the textbook. Assume that you would like to purchase a home with a list price of $399,000. Also assume that the home has a living area of 2500 square feet, it has 7 rooms, 3 bathrooms, is 15 years old, sits on a lot with an area of 0.87 acre, and the current owners pay $6000 in taxes. You would like to find the best regression or multiple regression equation so that your estimate of the selling price is the best estimate possible.

a. Which regression equation or multiple regression is best?

____

b. Explain your choice in part (a).

____

____

c. What is the best predicted value of the selling price? ______________

d. What is an advantage of knowing the best equation that can be used for predictions?

____

e. What is a disadvantage of not knowing the best equation that can be used for predictions?

____

10-12. ***Predicting Cholesterol Level*** Refer to the health exam results for females given in Data Set 1 in Appendix B of the textbook. The cholesterol level is the only variable requiring a blood sample. Assume that a clinic is not equipped to take blood samples. Can the cholesterol level be predicted using the other variables? Explain.

____

____

____

____

10-13. ***Manatee Deaths from Boats*** Listed below are the numbers of Florida manatee deaths related to encounters with watercraft (based on data from *The New York Times*). The data are listed in order, beginning with the year 1980 and ending with the year 2000.

16 24 20 15 34 33 33 39 43 50 47 53 38 35 49 42 60 54 67 82 78

Use the given data to find equations and coefficients of determination for the indicated models. Enter the results below.

	Equation	R^2
Linear	____________	______
Quadratic	____________	______
Logarithmic	____________	______
Exponential	____________	______
Power	____________	______

Based on the above results, which model appears to best fit the data? Why?

____

What is the best predicted value for 2001? In 2001, there were 82 watercraft–related manatee deaths. How does the predicted value compare to the actual value?

____

____

____

10-14. ***Activities with STATDISK: Exploring the Standard Error of Estimate***
Just as the standard deviation measures variation of data values away from their mean, the standard error of estimate measures variation of points away from their regression line. Specifically, the standard error of estimate, denoted by s_e, is a type of average spread of all the vertical distances between the observed and predicted values of the dependent variable.

This activity will help us understand what the standard error of estimate measures. Select **Datasets** and choose the Forecast and Actual Temperatures. We will use the actual high temperatures in column 1 for the dependent variable y, and we will use the data in each of columns 3, 5, and 7 for the independent variable x.

Select **Analysis**, then select **Correlation and Regression** for the following. (*Hint: Save results for Experiment 10-15.*)

Correlation Using Actual High Temperatures with Themselves
Using column 1 with *itself*, apply the **Correlation and Regression** module to find the following.

What is the standard error? __________

What is the linear correlation coefficient? __________

Describe the fit of the data points to the regression line.

____

Correlation Using Actual High Temperatures with 1 Day Predicted Highs
Using the 1 day predicted high temperatures in column 3 and the actual high temperatures in column 1, apply the **Correlation and Regression** module to find the following.

What is the standard error? __________

What is the linear correlation coefficient? __________

What is the total variation? ____________

What is the unexplained variation? ____________

Describe the fit of the data points to the regression line.

____

continued

Correlation Using Actual High Temperatures and 3 Day Predicted Highs
Using the 3 day predicted high temperatures in column 5 and the actual high temperatures in column 1, apply the **Correlation and Regression** module to find the following.

What is the standard error? __________

What is the linear correlation coefficient? __________

What is the total variation? ____________

What is the unexplained variation? __________

Describe the fit of the data points to the regression line.

____

Correlation Using Actual High Temperatures and 5 Day Predicted Highs
Using the 5 day predicted high temperatures in column 5 and the actual high temperatures in column 1, apply the **Correlation and Regression** module to find the following.

What is the standard error? __________

What is the linear correlation coefficient? __________

What is the total variation? ____________

What is the unexplained variation? __________

Describe the fit of the data points to the regression line.

____

Using the results of Correlation and Regression and your answers to the questions above, do the correlation coefficient and the standard error of estimate vary directly or indirectly?

____

Using the results of Correlation and Regression and your answers to the questions above, describe the connection between the standard error of estimate and the fit of the data points to the regression line.

____

____

continued

Why is the total variation constant throughout the four correlation analyses?

____

What is the relationship between the standard error and unexplained variation?

____

____

If a data set of 35 pairs of values has a standard error of estimate greater than 10, would you think that there would be a poor fit or good fit of the data points to the regression line? Explain.

____

____

10-15. ***Activities with STATDISK: Exploring Prediction Intervals*** Here is an analogy: A confidence interval is to a population parameter as a prediction interval is to an individual predicted value of the variable y. Specifically, a prediction interval allows us to estimate the true value of a predicted value. It is associated with a particular degree of confidence, and it is described with upper and lower limits. Instead of basing our prediction on a single point or value, we can base our prediction on an interval.

We can use STATDISK to find a prediction interval, but we must apply two of its modules and do some calculating. The textbook describes the prediction interval for an individual y at a given point x_0. Its format is $\hat{y} - E < y < \hat{y} + E$. Once you calculate $\hat{y}$ and E, you can easily obtain the prediction interval. First, we will evaluate E for $x_0 = 64$. Correlate columns 3 and 1. (Use the actual high temperatures for the dependent variable y, and use the one-day predicted high temperatures for the independent variable x. Find $t_{\alpha/2}$ from Table A-3 or by using STATDISK. Obtain the standard error from the Correlation and Regression results. To find the sum of squares of the independent variable, select **Data** and use **Descriptive Statistics** with the data in column 3. Substitute the values of $\bar{x}$, Σx, and Σx^2. When all the values are substituted, E can be evaluated as:

$$E = (2.037)\,(2.566954)\sqrt{1 + \frac{1}{35} + \frac{35(64 - 74.37143)^2}{35(194759) - (2603)^2}} = 5.5.$$

Since $\hat{y} = -4.337093 + 1.050249(64) = 62.9$, the prediction interval is

$$62.9 - 5.5 < y < 62.9 + 5.5$$
$$57.4 < y < 68.4.$$

continued

We are 95% confident that the actual temperature for a one-day forecast of 64°F will be a value between 57.4°F and 68.4°F.

The plot of the regression line is shown below. A sketch of the 95% prediction interval is drawn to show where the estimate of the actual values would be. In this case, we know the actual value since it is in the Y data set. The regression model is used to help you understand what a prediction interval tells you.

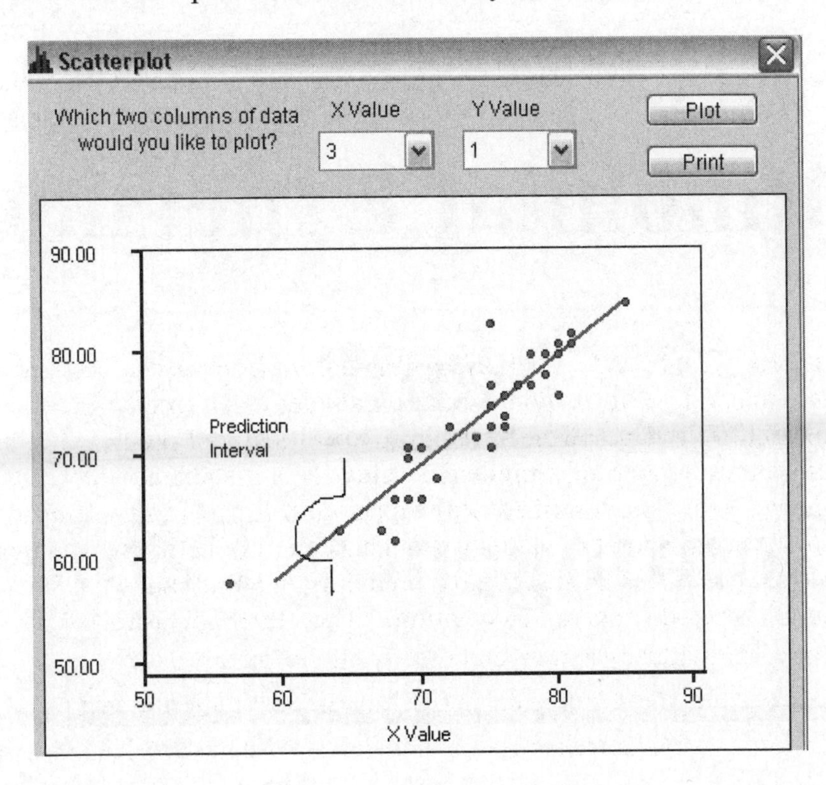

Assume that you need to predict a temperature based on a one-day forecast of 60°F. Find the predicted value and the prediction interval. Enter them below.

Predicted value: __________

Prediction interval: ____

11

Multinomial Experiments and Contingency Tables

11-1 Multinomial Experiments: Goodness-of-Fit

In Section 11-2 of the textbook, we deal with frequency counts from qualitative data that have been separated into different categories. The main objective is to determine whether the distribution of the sample data agrees with or "fits" some claimed distribution. Also, a multinomial experiment is defined as follows:

A **multinomial experiment** is an experiment that meets the following conditions.

1. The number of trials is fixed.
2. The trials are independent.
3. All outcomes of each trial must be classified into exactly one of several different categories.
4. The probabilities for the different categories remain constant for each trial.

STATDISK Procedure for Multinomial Experiments

1. Enter the *observed* frequencies in a column of the Statdisk data window. If the expected frequencies are not all the same, you must also enter a column of the *expected frequencies* or the *expected proportions* in a column of the Statdisk data window.
2. Select **Analysis** from the main menu.
3. Select **Multinomial Experiments** from the submenu.
4. You now are presented with the following two options:

   - Equal Expected Frequencies
   - Unequal Expected Frequencies

   If you want to test the claim that the different categories are all equally likely, select "Equal Expected Frequencies." If you want to test the claim that the different categories occur with some claimed proportions (not all equal), select the second item of "Unequal Expected Frequencies."

5. In the dialog box that now appears, enter a significance level, such as 0.05.
6. Select the column containing the observed frequencies and, if necessary, select the column containing the expected frequencies or the expected proportions.
7. Click on the **Evaluate** button.

8. Click on **Plot** to obtain a graph of the χ^2 distribution that includes the test statistic and critical value.

Section 11-2 in the textbook includes an example involving the analysis of leading digits on the amounts of checks from companies suspected of fraud. The objective is to determine whether the leading digits fit the frequencies found from Benford's Law, whereby the leading digit of 1 occurs 30.1% of the time, 2 occurs 17.6% of the time, and so on. (See Table 11–1 for the complete list of frequencies for Benford's Law.)

Table 11–1 Benford's Law: Distribution of Leading Digits

Leading Digit	1	2	3	4	5	6	7	8	9
Frequency According to Benford's Law	30.1%	17.6%	12.5%	9.7%	7.9%	6.7%	5.8%	5.1%	4.6%
Expected Leading Digits of 784 Checks Following Benford's Law	235.984	137.984	98.000	76.048	61.936	52.528	45.472	39.984	36.064
Observed Leading Digits of 784 Checks Analyzed for Fraud	0	15	0	76	479	183	8	23	0

As in Section 11–2 of the textbook, we will use a 0.01 significance level to test the claim that there is a discrepancy between the leading digits expected from Benford's Law and the leading digits observed on the 784 checks. Using the above STATDISK procedure, we begin by entering the observed and expected frequencies in two columns of the Statdisk data window. Then we select **Analysis**, then **Multinomial Experiments**, and we select the option of **Unequal Expected Frequencies**. Using a 0.01 significance level, we get the results shown below. We can see that the *P*-value of 0.0000 suggests that we reject the null hypothesis that the frequencies fit the distribution determined by Benford's Law. There is a discrepancy between the distribution of the leading digits observed on the checks and the distribution that follows Benford's Law. This is a strong indication that the check amounts are not the result of typical transactions.

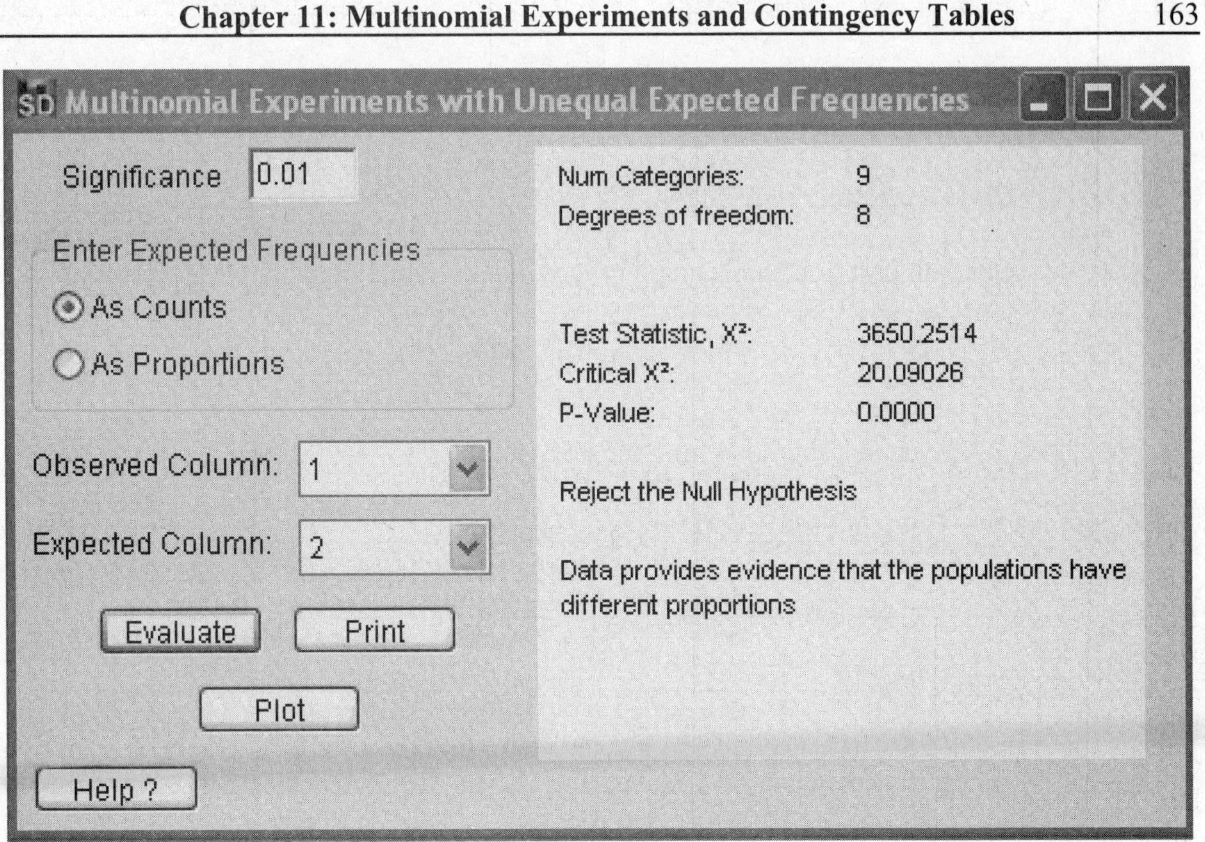

11-2 Contingency Tables

Section 11-3 of the textbook discusses contingency tables. Here is the STATDISK procedure for analyzing a contingency table. We can test for *independence* between the row and column variables, or we can conduct a test of homogeneity by testing that different populations have the same proportions of some characteristic.

STATDISK Procedure for Contingency Tables

1. Enter the observed frequencies in columns of the Statdisk data window. Enter the data in rows and columns as they appear in the contingency table.

2. Select **Analysis** from the main menu.

3. Select **Contingency Tables**.

4. In the dialog box that appears, enter a significance level such as 0.05 or 0.01.

5. Enter the columns that constitute the contingency table. For example, to enter

the frequencies from a 2 × 3 contingency table, use two rows and three columns.

6. Click on the **Evaluate** button.

7. Click on **Plot** to obtain a graph of the χ^2 distribution that includes the test statistic and critical value.

Consider the sample data in Table 11-4 from the textbook.

Table 11-4 Case-Control Study of Motorcycle Drivers

	Color of Helmet		
	Black	**White**	**Yellow/Orange**
Controls (not injured)	491	377	31
Cases (injured or killed)	213	112	8

We want to use a 0.05 significance level to test the claim that the color of the helmet worn by a motorcycle driver is independent of whether the driver is injured. Shown below are the STATDISK results.

```
Degrees of freedom: 2

Test Statistic, X²:    8.7747
Critical X²:           5.991471
P-Value:               0.0124

Reject the Null Hypothesis

Data provides evidence that the rows and
columns are related
```

We can see that the STATDISK display includes the important elements we need to make a decision. The *P*-value of 0.0124 is less than the significance level of 0.05, so we reject the null hypothesis of independence between the row and column variables. The test statistic and critical value are also provided in the display. (There is only one critical value, because all hypothesis tests of this type are right-tailed.) It appears that color of the helmet is not independent of whether the motorcycle is injured. This suggests that motorcycle drivers should consider the color of their helmets, because their safety could be affected.

11-3 Fisher Exact Test

A requirement for using the chi-square test with contingency tables is that every cell must have an expected frequency of at least 5. If a contingency table does have an expected frequency less than 5, STATDISK's **Contingency Table** function will yield an error message and will not provide test results. If a 2 × 2 table has an expected frequency less than 5, the Fisher exact test can be used instead of the chi-square test. The Fisher exact test provides an *exact P*-value and does not require an approximation technique. STATDISK includes a new function for the Fisher exact test, and the procedure is described here.

STATDISK Procedure for the Fisher Exact Test

1. Select **Analysis** from the main menu.

2. Select **Fisher Exact Test**.

3. In the dialog box that appears, enter the four frequencies in the cells of the table. (Enter a frequency, press the **Tab** key, then enter another frequency, and so on. Do not try to enter row and column totals; they will be provided by STATDISK.)

4. Click on the **Evaluate** button.

Consider the sample data in Table 11-6 from the textbook. The expected frequencies are shown in parentheses, and we can see that one of them is less than 5.

Table 11-6 Helmets and Facial Injuries in Bicycle Accidents
Expected frequencies are in parentheses.

	Helmet Worn	No Helmet
Facial injuries received	2 (3)	13 (12)
All injuries nonfacial	6 (5)	19 (20)

The STATDISK display is shown on the following page. The exact *P*-value of 0.6857 is not small (such as less than 0.05), so we fail to reject the null hypothesis that wearing a helmet and receiving facial injuries are independent.

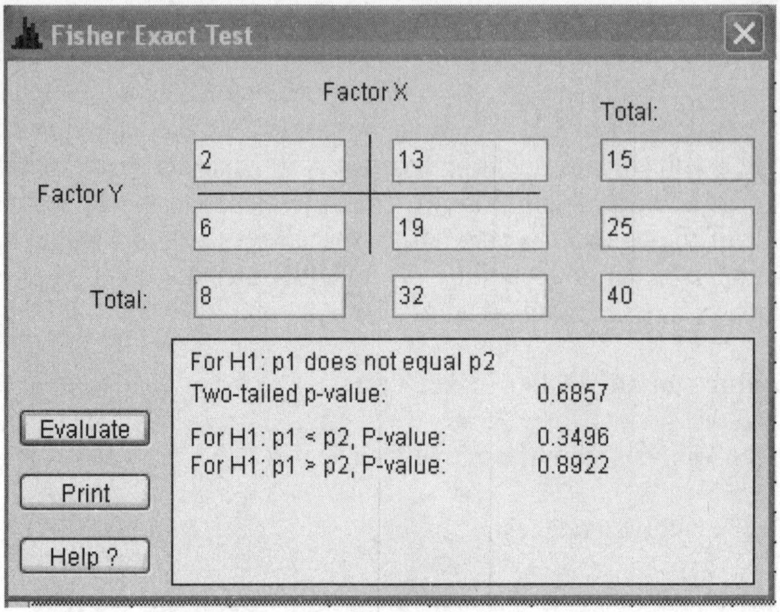

Fisher Exact Test

Factor X

Total:

Factor Y

			Total:
Factor Y	2	13	15
	6	19	25
Total:	8	32	40

For H1: p1 does not equal p2
Two-tailed p-value: 0.6857

For H1: p1 < p2, P-value: 0.3496
For H1: p1 > p2, P-value: 0.8922

Evaluate
Print
Help ?

11-4 McNemar's Test for Matched Pairs

The contingency table procedures in Section 11-3 of the textbook are based on *independent* data. For 2 × 2 tables consisting of frequency counts that result from *matched pairs*, we do not have independence and, for such cases, we can use McNemar's test for matched pairs. We can use McNemar's test for the null hypothesis that frequencies from the discordant (different) categories occur in the same proportion. Here is the STATDISK procedure.

STATDISK Procedure for McNemar's Test

1. Select **Analysis** from the main menu.

2. Select **McNemar's Test**.

3. In the dialog box that appears, enter the four frequencies in the cells of the table. (Enter a frequency, press the **Tab** key, then enter another frequency, and so on. Do not try to enter row and column totals; they will be provided by STATDISK.)

4. Enter a significance level, such as 0.05.

5. Click on the **Evaluate** button.

Consider the matched data form 80 subjects, with each subject treated with Fungacream on one foot while the other foot was treated with Pedacream. The sample results are in the table below, and the STATDISK results follow the table.

Table 11-8 Clinical Trials of Treatments for Athlete's Foot

		Treatment with Pedacream	
		Cured	Not cured
Treatment with Fungacream	Cured	12	8
	Not cured	40	20

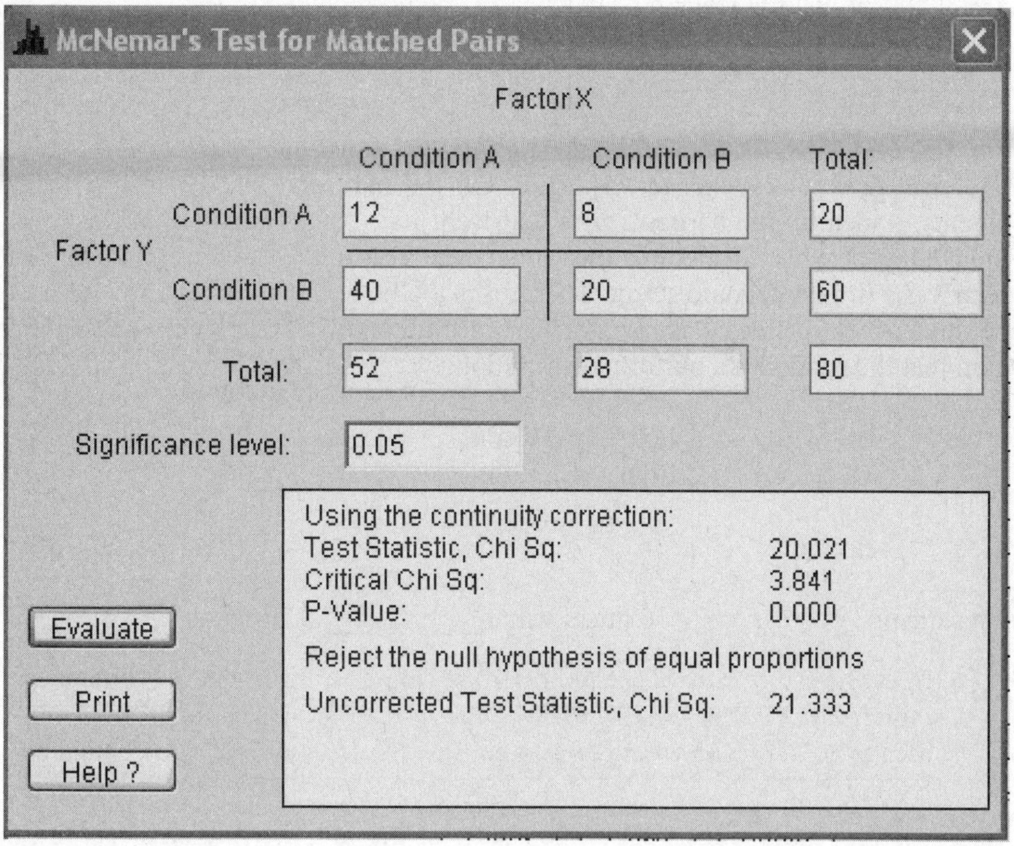

McNemar's Test for Matched Pairs

		Factor X		
		Condition A	Condition B	Total:
Factor Y	Condition A	12	8	20
	Condition B	40	20	60
	Total:	52	28	80

Significance level: 0.05

Using the continuity correction:
Test Statistic, Chi Sq: 20.021
Critical Chi Sq: 3.841
P-Value: 0.000
Reject the null hypothesis of equal proportions
Uncorrected Test Statistic, Chi Sq: 21.333

Evaluate

Print

Help ?

The STATDISK results include the chi-square test statistic, the critical value, and the *P*-value. Because the *P*-value of 0.000 is small (less than the significance level of 0.05), we reject the null hypothesis of equal proportions. We conclude that the two treatments are not equally effective. Analyzing the frequencies of 8 and 40, we see that many more feet were cured with Pedacream than Fungacream, so the Pedacream treatment appears to be more effective.

CHAPTER 11: Multinomial Experiments and Contingency Tables

11-1. ***Loaded Die*** The author drilled a hole in a die and filled it with a lead weight, then proceeded to roll it 200 times. Here are the observed frequencies for the outcomes of 1, 2, 3, 4, 5, and 6 respectively: 27, 31, 42, 40, 28, 32. Use a 0.05 significance level to test the claim that the outcomes are not equally likely.

Test statistic:__________ Critical value:__________ *P*-value:__________

Conclusion:____

____

Does it appear that the loaded die behaves differently than a fair die?

____

11–2. ***Flat Tire and Missed Class*** A classic tale involves four car-pooling students who missed a test and gave as an excuse a flat tire. On the makeup test, the instructor asked the students to identify the particular tire that went flat. If they really didn't have a flat tire, would they be able to identify the same tire? The author asked 41 other students to identify the tire they would select. The results are listed in the following table (except for one student who selected the spare). Use a 0.05 significance level to test the author's claim that the results fit a uniform distribution.

Tire	Left front	Right front	Left rear	Right rear
Number selected	11	15	8	6

Test statistic:__________ Critical value:__________ *P*-value:__________

Conclusion:____

____

What does the result suggest about the ability of the four students to select the same tire when they really didn't have a flat?

____

11–3. ***Grade and Seating Location*** Do "A" students tend to sit in a particular part of the classroom? The author recorded the locations of the students who received grades of A, with these results: 17 sat in the front, 9 sat in the middle, and 5 sat in the back of the classroom. Is there sufficient evidence to support the claim that the "A" students are not evenly distributed throughout the classroom? If so, does that mean you can increase your likelihood of getting an A by sitting in the front?

Test statistic:__________ Critical value:__________ P-value:__________

Conclusion:____

____

11–4. ***Post Position and Winning Horse Races*** Many people believe that when a horse races, it has a better chance of winning if its starting line-up position is closer to the rail on the inside of the track. The starting position of 1 is closest to the inside rail, followed by position 2, and so on. The accompanying table lists the numbers of wins for horses in the different starting positions (based on data from the *New York Post*). Test the claim that the probabilities of winning in the different post positions are not all the same.

Starting Position	1	2	3	4	5	6	7	8
Number of wins	29	19	18	25	17	10	15	11

Test statistic:__________ Critical value:__________ P-value:__________

Conclusion:____

____

11–5. ***Measuring Pulse Rates*** According to one procedure used for analyzing data, when certain quantities are measured, the last digits tend to be uniformly distributed, but if they are estimated or reported, the last digits tend to have disproportionately more 0s or 5s. Refer to Data Set 1 in Appendix B of the textbook and use the last digits of the pulse rates of the 80 men and women. Those pulse rates were obtained as part of the National Health and Examination Survey. Test the claim that the last digits of 0, 1, 2, . . . , 9 occur with the same frequency.

Test statistic:__________ Critical value:__________ P-value:__________

Conclusion:____

____

Based on the observed digits, what can be inferred about the procedure used to obtain the pulse rates?

____

11–6. ***M&M Candies*** Mars, Inc. claims that its M&M plain candies are distributed with the following color percentages: 16% green, 20% orange, 14% yellow, 24% blue, 13% red, and 13% brown. Refer to Data Set 13 in Appendix B and use the sample data to test the claim that the color distribution is as claimed by Mars, Inc. Use a 0.05 significance level.

Test statistic:__________ Critical value:__________ *P*-value:__________

Conclusion:____

____

11–7. ***Author's Check Amounts and Benford's Law*** The table below lists the observed frequencies of the leading digits from the amounts of the last 200 checks that the author wrote. Using a 0.05 significance level, test the claim that they come from a population of leading digits that conform to Benford's Law. (See the first two rows of Table 11–1 included in Section 11–1 of this manual/workbook.)

Leading Digit	1	2	3	4	5	6	7	8	9
Frequency	72	23	26	20	21	18	8	8	4

Test statistic:__________ Critical value:__________ *P*-value:__________

Conclusion:____

____

11–8. ***Do World War II Bomb Hits Fit a Poisson Distribution?*** In analyzing hits by V-1 buzz bombs in World War II, South London was subdivided into regions, each with an area of 0.25 km^2. Use the values listed here and test the claim that the actual frequencies fit a Poisson distribution. Use a 0.05 significance level.

Number of bomb hits	0	1	2	3	4 or more
Actual number of regions	229	211	93	35	8
Expected number of regions (from Poisson distribution)	227.5	211.4	97.9	30.5	8.7

Test statistic:__________ Critical value:__________ *P*-value:__________

Conclusion:____

____

11-9. ***Testing a Normal Distribution*** In this experiment we will use STATDISK's ability to generate normally distributed random numbers. We will then test the sample data to determine if they actually do fit a normal distribution.

a. Generate 1000 random numbers from a normal distribution with a mean of 100 and a standard deviation of 15. (IQ scores have these parameters.) Select **Data**, then **Normal Generator**.

b. Arrange the generated data in order. (Use **Copy/Paste** to copy the data to the STATDISK data window, where the **Data tools** button gives you the option of sorting a column of data.)

c. Examine the sorted list and determine the frequency for each of the categories listed below. Enter those frequencies in the spaces provided. (The expected frequencies were found using the methods of Chapter 6 in the textbook.)

	Observed Frequency	Expected Frequency
Below 55:	__________	1
55-70:	__________	22
70-85:	__________	136
85-100:	__________	341
100-115:	__________	341
115-130:	__________	136
130-145:	__________	22
Above 145:	__________	1

d. Use STATDISK to test the claim that the randomly generated numbers actually do fit a normal distribution with mean 100 and standard deviation 15.

Test statistic:__________ Critical value:__________ *P*-value:__________

Conclusion:____

____

11-10. ***E–Mail and Privacy*** Workers and senior–level bosses were asked if it was seriously unethical to monitor employee e–mail, and the results are summarized in the table (based on data from a Gallup poll). Use a 0.05 significance level to test the claim that the response is independent of whether the subject is a worker or a senior–level boss.

	Yes	No
Workers	192	244
Bosses	40	81

Test statistic:__________ Critical value:__________ *P*-value:__________

Conclusion:____

____

Does the conclusion change if a significance level of 0.01 is used instead of 0.05? Do workers and bosses appear to agree on this issue?

____

11–11. ***Accuracy of Polygraph Tests*** The data in the accompanying table summarize results from tests of the accuracy of polygraphs (based on data from the Office of Technology Assessment). Use a 0.05 significance level to test the claim that whether the subject lies is independent of the polygraph indication.

	Polygraph Indicated Truth	Polygraph Indicated Lie
Subject actually told the truth	65	15
Subject actually told a lie	3	17

Test statistic:__________ Critical value:__________ *P*-value:__________

Conclusion:____

____

What do the results suggest about the effectiveness of polygraphs?

____

11–12. ***Fear of Flying Gender Gap*** The Marist Institute for Public Opinion conducted a poll of 1014 adults, 48% of whom were men. The poll results show that 12% of the men and 33% of the women fear flying. After constructing a contingency table that summarizes the data in the form of frequency counts, use a 0.05 significance level to test the claim that gender is independent of the fear of flying.

Test statistic:__________ Critical value:__________ *P*-value:__________

Conclusion:____

____

11–13. ***Occupational Hazards*** Use the data in the table to test the claim that occupation is independent of whether the cause of death was homicide. The table is based on data from the U.S. Department of Labor, Bureau of Labor Statistics.

	Police	Cashiers	Taxi Drivers	Guards
Homicide	82	107	70	59
Cause of Death Other than Homicide	92	9	29	42

Test statistic:__________ Critical value:__________ *P*-value:__________

Conclusion:____

____

Does any particular occupation appear to be most prone to homicides? If so, which one?

____

How are the results affected if the order of the rows is switched?

____

How are the results affected by the presence of an outlier? If we change the first entry from 82 to 8200, are the results dramatically affected?

____

11–14. ***Treating Athlete's Foot*** Assume that subjects are inflicted with athlete's foot on each of their feet. Also assume that for each subject, one foot is treated with a fungicide solution while the other foot is given a placebo. The results are given in the accompanying table. Using a 0.05 significance level, test the effectiveness of the treatment.

		Fungicide Treatment	
		Cure	No cure
Placebo	Cure	5	12
	No cure	22	55

Test statistic:__________ Critical value:__________ *P*-value:__________

Conclusion:____

____

11-15. ***Treating Athlete's Foot*** Repeat Experiment 11-14 after changing the frequency of 22 to 66.

Test statistic:__________ Critical value:__________ *P*-value:__________

Conclusion:____

____

11-16. ***PET/CT Compared to MRI*** In the article "Whole-Body Dual-Modality PET/CT and Whole Body MRI for Tumor Staging in Oncology" (by Antoch et al, *Journal of the American Medical Association*, Vol. 290, No. 24), the authors cite the importance of accurately identifying the stage of a tumor. Accurate staging is critical for determining appropriate therapy. The article discusses a study involving the accuracy of positron emission tomography (PET) and computed tomography (CT) compared to magnetic resonance imaging (MRI). Using the data in the table for 50 tumors analyzed with both technologies, does there appear to be a difference in accuracy? Does either technology appear to be better? Enter the results on the following page.

		PET/CT	
		Correct	Incorrect
MRI	Correct	36	1
	Incorrect	11	2

Test statistic:__________ Critical value:__________ *P*-value:__________

Conclusion:____

____

11-17. ***Testing a Treatment*** In the article "Eradication of Small Intestinal Bacterial Overgrowth Reduces Symptoms of Irritable Bowel Syndrome" (by Pimentel, Chow, Lin, *American Journal of Gastroenterology*, Vol. 95, No. 12), the authors include a discussion of whether antibiotic treatment of bacteria overgrowth reduces intestinal complaints. McNemar's test was used to analyze results for those subjects with eradication of bacterial overgrowth. Using the data in the given table, does the treatment appear to be effective against abdominal pain?

		Abdominal pain before treatment?	
		Yes	No
Abdominal pain after treatment?	Yes	11	1
	No	14	3

Test statistic:__________ Critical value:__________ *P*-value:__________

Conclusion:____

____

Analysis of Variance

12-1 One-Way Analysis of Variance

One–way analysis of variance is used to test the claim that three or more populations have the same mean. When the textbook discusses one-way analysis of variance, it is noted that the term "one-way" is used because the sample data are separated into groups according to one characteristic or "factor". For example, the textbook discusses the data in the table that is reproduced below as Table 12–1. Here is the key issue: Do the different treatment categories result in weights of trees having means that are not all the same? In this case, the factor is the treatment.

Table 12–1 **Weights (kg) of Poplar Trees**

	Treatment			
	None	Fertilizer	Irrigation	Fertilizer and Irrigation
	0.15	1.34	0.23	2.03
	0.02	0.14	0.04	0.27
	0.16	0.02	0.34	0.92
	0.37	0.08	0.16	1.07
	0.22	0.08	0.05	2.38
n	5	5	5	5
$\bar{x}$	0.184	0.332	0.164	1.334
s	0.127	0.565	0.126	0.859

The following procedure describes how STATDISK can be used with such collections of sample data to test the claim that the different samples come from populations with the same mean. For the weights of the poplar trees in Table 12-1, the claim of equal means leads to these hypotheses:

H_0: $\mu_1 = \mu_2 = \mu_3 = \mu_4$

H_1: At least one of the four population means is different from the others.

STATDISK Procedure for One-Way Analysis of Variance

1. Enter the data in separate columns of the Statdisk data window.

2. Select **Analysis** from the main menu.

3. Select **One-Way Analysis of Variance** from the submenu.

4. In the dialog box, enter a significance level, such as 0.05 or 0.01.

5. Select the columns to be used for the analysis of variance. If a box already has a check mark and you do not want to include it, click on the box to remove the check mark. If a box does not have a check mark and you want to include it, click on the box to make the check mark appear.

6. Click on the **Evaluate** button.

7. Click on **Plot** to obtain a graph that includes the critical value and test statistic.

If you use the above steps with the data in Table 12-1, the STATDISK result will appear as follows.

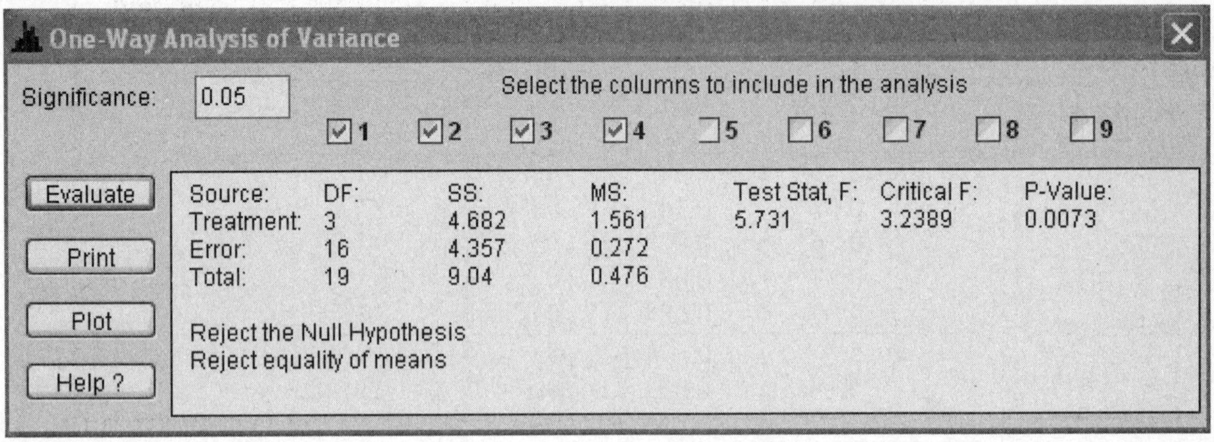

One-Way Analysis of Variance

Significance: 0.05

Select the columns to include in the analysis

☑1 ☑2 ☑3 ☑4 ☐5 ☐6 ☐7 ☐8 ☐9

Evaluate
Print
Plot
Help ?

Source:	DF:	SS:	MS:	Test Stat, F:	Critical F:	P-Value:
Treatment:	3	4.682	1.561	5.731	3.2389	0.0073
Error:	16	4.357	0.272			
Total:	19	9.04	0.476			

Reject the Null Hypothesis
Reject equality of means

The *P*-value of 0.0073 indicates that there is sufficient sample evidence to warrant rejection of the null hypothesis that $\mu_1 = \mu_2 = \mu_3 = \mu_4$. The test statistic of $F = 5.731$ is also provided along with the critical value of $F = 3.2389$. The values of the SS and MS components are also provided. If you click on the **Plot** button, you will get a graph showing the test statistic and critical value.

Caution: It is easy to feed STATDISK (or any other software package) data that can be processed quickly and painlessly, but we should *think* about what we are doing. We should consider the assumptions for the test being used, and we should *explore* the data before jumping into a formal procedure such as analysis of variance. Carefully explore the important characteristics of data, including the center (through means and medians), variation (through standard deviations and ranges), distribution (through histograms and boxplots), outliers, and any changing patterns over time.

12-2 Two-Way Analysis of Variance

Two-way analysis of variance involves *two* factors, such as site and treatment in Table 12-4 from Section 12–3 in *Elementary Statistics*. The two–way analysis of variance procedure requires that we test for (1) an interaction effect between the two factors; (2) an effect from the row factor; (3) an effect from the column factor.

Table 12-4 Poplar Tree Weights (kg)

	No Treatment	Fertilizer	Irrigation	Fertilizer and Irrigation
	0.15	1.34	0.23	2.03
Site 1	0.02	0.14	0.04	0.27
(rich, moist)	0.16	0.02	0.34	0.92
	0.37	0.08	0.16	1.07
	0.22	0.08	0.05	2.38
	0.60	1.16	0.65	0.22
Site 2	1.11	0.93	0.08	2.13
(sandy, dry)	0.07	0.30	0.62	2.33
	0.07	0.59	0.01	1.74
	0.44	0.17	0.03	0.12

STATDISK Procedure for Two-Way Analysis of Variance

1. Select **Analysis** from the main menu.

2. Select **Two-Way Analysis of Variance** from the submenu.

3. In the dialog box, enter the significance level, such as 0.05 or 0.01.

4. In the dialog box, enter the number of categories for the row variable, enter the number of categories for the column variable, and enter the number of values in each cell. For the sample data in Table 12-4, enter 2 for the number of categories for the row variable (site 1/site 2), enter 4 for the number of categories of the column variable (for the 4 different treatments), and enter 5 for the number of values in each cell. Click on **Continue** when finished.

5. STATDISK will automatically generate a format for entering the sample data. You will be given row and column numbers, so enter the sample values according to their locations See the STATDISK display below and see how the sample values from Table 12-4 are entered.

6. Click **Evaluate** after all sample values have been entered.

See Section 12–3 in *Elementary Statistics* for the basic procedure for two–way analysis of variance, and note that it involves three distinct components: (1) Test for an interaction between the two variables; (2) test for an effect from the row variable; (3) test for an effect from the column variable. In the results included below, see that the interaction is associated with a *P*-value of 0.9154, so we fail to reject the null hypothesis of no interaction effect. That is, there does not appear to be an effect due to an interaction between site and treatment. For the row variable (site), the *P*-value is 0.3742, so there does not appear to be an effect from the site location. For the column variable (treatment), the *P*-value of 0.0006 suggests that there is an effect from treatment. That is, the poplar tree weights appear to be affected by the treatments that they are given.

Two-Way Analysis of Variance

Significance: 0.05

Number of categories for ROW variable: 2

Number of categories for COLUMN variable: 4

Number of values in each cell: 5

Continue

Row	Column	Value
1	1	0.15
1	1	0.02
1	1	0.16
1	1	0.37
1	1	0.22
1	2	1.34
1	2	0.14
1	2	0.02
1	2	0.08
1	2	0.08
1	3	0.23
1	3	0.04
1	3	0.34
1	3	0.16
1	3	0.05
1	4	2.03

Evaluate Print Paste Clear

Source:	DF:	SS:	MS:	Test Stat, F:	Critical F:	P-Value:
Interaction:	3	0.1716	0.0572	0.17	2.9011	0.9154
Row Variable:	1	0.2722	0.2722	0.8122	4.1491	0.3742
Column Variable:	3	7.547	2.5157	7.5048	2.9011	0.0006

Help ?

Special Case: One Observation per Cell and No Interaction The textbook describes the special case consisting of data with only one observation per cell. If there is only one observation per cell, and if it seems reasonable to assume (based on knowledge about the circumstances) that there is no interaction between the two factors, make that assumption and then proceed as before to test effects from the row and column factors separately. STATDISK does work with this special case of only one observation per cell. Simply follow the same procedure given in this section.

CHAPTER 12 EXPERIMENTS: Analysis of Variance

In Experiments 12–1 and 12–2, use the listed sample data from car crash experiments conducted by the National Transportation Safety Administration. New cars were purchased and crashed into a fixed barrier at 35 mi/h, and the listed measurements were recorded for the dummy in the driver's seat. The subcompact cars are the Ford Escort, Honda Civic, Hyundai Accent, Nissan Sentra, and Saturn SL4. The compact cars are Chevrolet Cavalier, Dodge Neon, Mazda 626 DX, Pontiac Sunfire, and Subaru Legacy. The Midsize cars are Chevrolet Camaro, Dodge Intrepid, Ford Mustang, Honda Accord, and Volvo S70. The full–size cars are Audi A8, Cadillac Deville, Ford Crown Victoria, Oldsmobile Aurora, and Pontiac Bonneville.

12–1. ***Head Injury in a Car Crash*** The head injury data (in hic) are given below. Use a 0.05 significance level to test the null hypothesis that the different weight categories have the same mean. Do the data suggest that larger cars are safer?

Subcompact:	681	428	917	898	420
Compact:	643	655	442	514	525
Midsize:	469	727	525	454	259
Full–Size:	384	656	602	687	360

SS(treatment): _______ MS(treatment): _______ Test statistic F: _______

SS(error): _______ MS(error): _______ P-value: _______

SS(total): _______

Conclusion:____

12–2. ***Chest Deceleration in a Car Crash*** The chest deceleration data (g) are given below. Use a 0.05 significance level to test the null hypothesis that the different weight categories have the same mean. Do the data suggest that larger cars are safer?

Subcompact:	55	47	59	49	42
Compact:	57	57	46	54	51
Midsize:	45	53	49	51	46
Full–Size:	44	45	39	58	44

SS(treatment): _______ MS(treatment): _______ Test statistic F: _______

SS(error): _______ MS(error): _______ P-value: _______

SS(total): _______

Conclusion:____

12–3. ***Archeology: Skull Breadths from Different Epochs*** The values in the table are measured maximum breadths of male Egyptian skulls from different epochs (based on data from Ancient Races of the Thebaid, by Thomson and Randall-Maciver). Changes in head shape over time suggest that interbreeding occurred with immigrant populations. Use a 0.05 significance level to test the claim that the different epochs do not all have the same mean.

4000 B.C.	1850 B.C.	150 A.D.
131	129	128
138	134	138
125	136	136
129	137	139
132	137	141
135	129	142
132	136	137
134	138	145
138	134	137

SS(treatment): _______ MS(treatment): _______ Test statistic *F*: _______

SS(error): _______ MS(error): _______ *P*-value: _______

SS(total): _______

Conclusion:____

____

12–4. ***Mean Weights of M&Ms*** Refer to Data Set 13 in Appendix B from the textbook. At the 0.05 significance level, test the claim that the mean weight of M&Ms is the same for each of the six different color populations.

SS(treatment): _______ MS(treatment): _______ Test statistic *F*: _______

SS(error): _______ MS(error): _______ *P*-value: _______

SS(total): _______

Conclusion:____

If it is the intent of Mars, Inc., to make the candies so that the different color populations have the same mean weight, do these results suggest that the company has a problem requiring corrective action?

____

12–5. ***Homerun Distances*** Refer to Data Set 17 in Appendix B from the textbook. Use a 0.05 significance level to test the claim that the homeruns hit by Barry Bonds, Mark McGwire, and Sammy Sosa have mean distances that are not all the same.

SS(treatment): ________ MS(treatment): ________ Test statistic F: ________

SS(error): ________ MS(error): ________ P-value: ________

SS(total): ________

Conclusion:____

____

Do the homerun distances explain the fact that as of this writing, Barry Bonds has the most homeruns in one season, while Mark McGwire has the second highest number of runs?

____

12–6. ***Simulations*** Use STATDISK to randomly generate three different samples of 500 values each. (Select **Data**, then select **Normal Generator**.) For the first two samples, use a normal distribution with a mean of 100 and a standard deviation of 15. For the third sample, use a normal distribution with a mean of 105 and a standard deviation of 15. We know that the three populations have different means, but do the methods of analysis of variance allow you to conclude that the means are different? Explain.

____

____

12–7. ***Pulse Rates*** The following table lists pulse rates from Data Set 1 in Appendix B.

	Age		
	Under 20	20–40	Over 40
Male	96 64 68 60	64 88 72 64	68 72 60 88
Female	76 64 76 68	72 88 72 68	60 68 72 64

Test for an interaction between gender and age.

Test statistic: ________ Critical value:________ P-value:________

Conclusion:____

____

(*continued*)

Test the claim that pulse rates are affected by gender.

Test statistic: _________ Critical value:_________ *P*-value:_________

Conclusion:____

Test the claim that pulse rates are affected by age.

Test statistic: _________ Critical value:_________ *P*-value:_________

Conclusion:____

12–8. ***Marathon Times*** Listed below are New York Marathon running times (in seconds) for randomly selected runners who completed the marathon.

	Age		
	21–29	30–39	40 and over
Male	13,615	14,677	14,528
	18,784	16,090	17,034
	14,256	14,086	14,935
	10,905	16,461	14,996
	12,077	20,808	22,146
Female	16,401	15,357	17,260
	14,216	16,771	25,399
	15,402	15,036	18,647
	15,326	16,297	15,077
	12,047	17,636	25,898

(*continued*)

Test for an interaction between gender and age.

Test statistic: __________ Critical value:__________ *P*-value:__________

Conclusion:____

____

Test the claim that running times are affected by gender.

Test statistic: __________ Critical value:__________ *P*-value:__________

Conclusion:____

____

Test the claim that running times are affected by age.

Test statistic: __________ Critical value:__________ *P*-value:__________

Conclusion:____

____

Nonparametric Statistics

13-1 Nonparametric Methods

STATDISK includes a wide variety of nonparametric procedures. It includes procedures for all of the nonparametric methods described in the Triola textbooks that include a chapter of Nonparametric Statistics. The sections of this chapter correspond to those in the textbook.

Section 13-1 in the textbook introduces some basic principles of nonparametric methods. The textbook notes that prior to Chapter 13, most of the methods of inferential statistics are called *parametric* methods, because they are based on sampling from a population with specific parameters (such as the mean μ, standard deviation σ, or proportion p). Those parametric methods usually have some fairly strict conditions, such as a requirement that the sample data must come from a normally distributed population. Because nonparametric methods do not require specific distributions (such as the normal distribution), these nonparametric methods are often called **distribution-free tests**. The following sections describe some of the more important and commonly-used nonparametric methods.

13-2 Sign Test

Section 13-2 in *Elementary Statistics* includes the following definition.

Definition
The **sign test** is a nonparametric (distribution-free) test that uses plus and minus signs to test different claims, including these:

1. Given matched pairs of sample data, test claims about the medians of the two populations.
2. Given nominal data, test claims about the proportion of some category.
3. Test claims about the median of a single population.

STATDISK makes it possible to work with all three of the above cases. We will first describe the STATDISK procedure, then we will illustrate it with an example.

STATDISK Procedure for the Sign Test

1. Either determine the number of positive signs and the number of negative signs, or enter the paired data in columns of the Statdisk data window.

2. Select **Analysis** from the main menu.

3. Select **Sign Tests** from the submenu.

4. If you already know the number of positive and negative signs (as in cases involving nominal data), select the option of **Given Number of Signs**.

   If you have sample *paired* data, select the option of **Given Pairs of Values**.

5. The content of the dialog box will depend on the choice made in step 3. Both cases require that you select the form of the claim being tested and a significance

level such as 0.05 or 0.01. You must then enter the numbers of positive and negative signs, or the columns containing the original pairs of data.

6. Click on the **Evaluate** button.

7. Click on **Plot** to obtain a graph that includes the test statistic and critical value. The plot will be generated only if the normal approximation is used (because $n > 25$).

Section 13-2 includes the data in Table 13-3, which is reproduced below. William Gosset published the article "The Probable Error of a Mean" under the pseudonym of "Student" (*Biometrika*, Vol. 6, No. 1). He included the data listed in Table 13–3 for two different types of corn seed (regular and kiln dried) that were used on *adjacent* plots of land. The listed values are the yields of head corn in pounds per acre. We want to test the claim that the yield is the same for the two different types of seed. The sample values were entered in two columns of the STATDISK data window, and the STATDISK sign test results are shown below the table. The results show that the test statistic is $x = 4$, the critical value is $x = 1$, and we fail to reject the null hypothesis. There is not sufficient sample evidence to warrant rejection of the claim that the two types of seed produce the same yield.

Table 13–3 Yields of Corn from Different Seeds

Regular	1903	1935	1910	2496	2108	1961	2060	1444	1612	1316	1511
Kiln Dried	2009	1915	2011	2463	2180	1925	2122	1482	1542	1443	1535

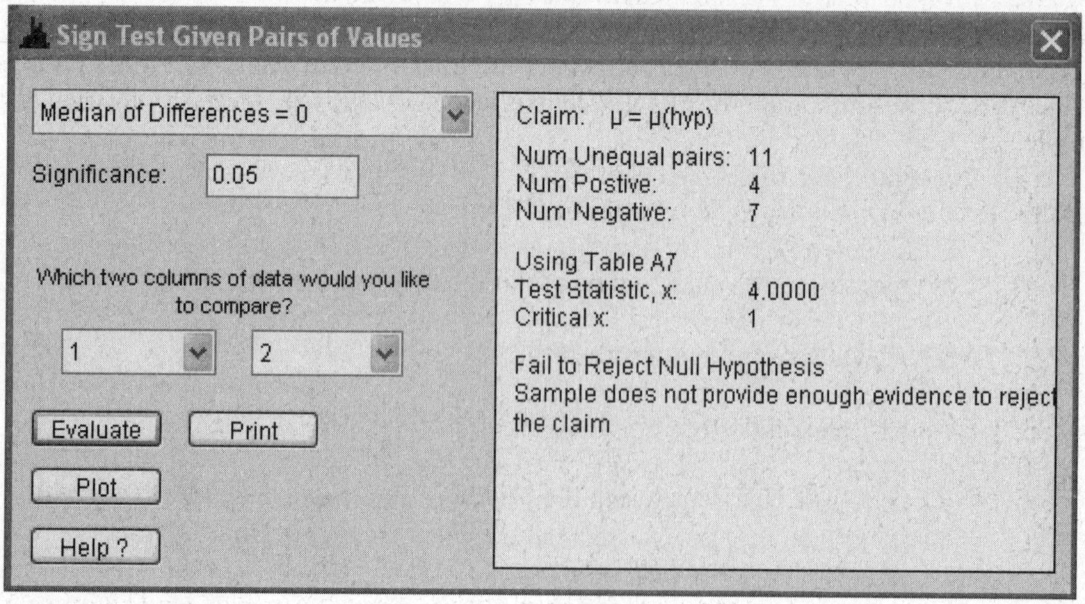

Sign Test Given Pairs of Values

Median of Differences = 0

Significance: 0.05

Which two columns of data would you like to compare?

1 2

Evaluate Print

Plot

Help ?

Claim: μ = μ(hyp)

Num Unequal pairs: 11
Num Postive: 4
Num Negative: 7

Using Table A7
Test Statistic, x: 4.0000
Critical x: 1

Fail to Reject Null Hypothesis
Sample does not provide enough evidence to reject the claim

The preceding example uses the original paired sample data, but STATDISK does allow you to enter the number of positive signs and the number of negative signs if those numbers are know. The textbook includes an example involving 325 births, with 295 resulting in girls (so there are 295 girls and 30 boys). Using STATDISK, we could select the option of **Given Number of Signs**, and we could enter 295 and 30 for the numbers of signs. (In fact, you might often find it easier to manually determine the signs than to enter the original paired data.)

13-3 Wilcoxon Signed-Ranks Test

Section 13-3 in *Elementary Statistics* describes the Wilcoxon signed-ranks test, and the following definition is given.

Definition
The **Wilcoxon signed-ranks test** is a nonparametric test that uses ranks of sample data consisting of *matched pairs*. It is used to test the null hypothesis that the population of differences has a median equal to 0. The null and alternative hypotheses are as follows.

H_0: The matched pairs have differences that come from a population with a median equal to zero.
H_1: The matched pairs have differences that come from a population with a nonzero median.

The textbook makes this important point: The Wilcoxon signed-ranks test and the sign test can both be used with sample data consisting of matched pairs, but the sign test uses only the signs of the differences and not their actual magnitudes (how large the numbers are). The Wilcoxon signed-ranks test uses ranks, so the magnitudes of the differences are taken into account. Because the Wilcoxon signed-ranks test incorporates and uses more information than the sign test, it tends to yield conclusions that better reflect the true nature of the data. First we describe the STATDISK procedure for conducting a Wilcoxon signed-ranks test, then we illustrate it with an example.

STATDISK Procedure for the Wilcoxon Signed-Ranks Test

1. Enter the paired data in columns of the Statdisk data window.

2. Select **Analysis** from the main menu.

3. Select **Wilcoxon Tests** from the submenu.

4. You must now choose between the following two options.

   Signed-Ranks Test (Select this option with *matched pairs*.)
   Rank-Sum Test (Select this option with two *independent* samples)

5. After selecting the **Signed-Ranks Test** option, proceed to enter a significance level, such as 0.05 or 0.01.

6. Select the columns of the Statdisk data window that contain the paired data.

7. Click on the **Evaluate** button.

8. Click on the **Plot** button to see a graph that includes the test statistic and critical values. The graph will be displayed only if the normal approximation is used (because $n > 30$).

Section 13-3 of the textbook includes an example that uses the same sample data listed in Table 13-3, shown in the preceding example of this manual/workbook. Using the same sample data from Table 13–3 with the Wilcoxon signed–ranks test, we get the STATDISK display shown below. Based on this display, we see that the test statistic is $T = 15$, the critical value is $T = 11$, and the conclusion is failure to reject the null hypothesis. There is not sufficient evidence to warrant rejection of the null hypothesis that the differences are from a population with a median equal to 0. It appears that the differences have a median of 0, so there is no significant difference between the yields from the two different types of seed.

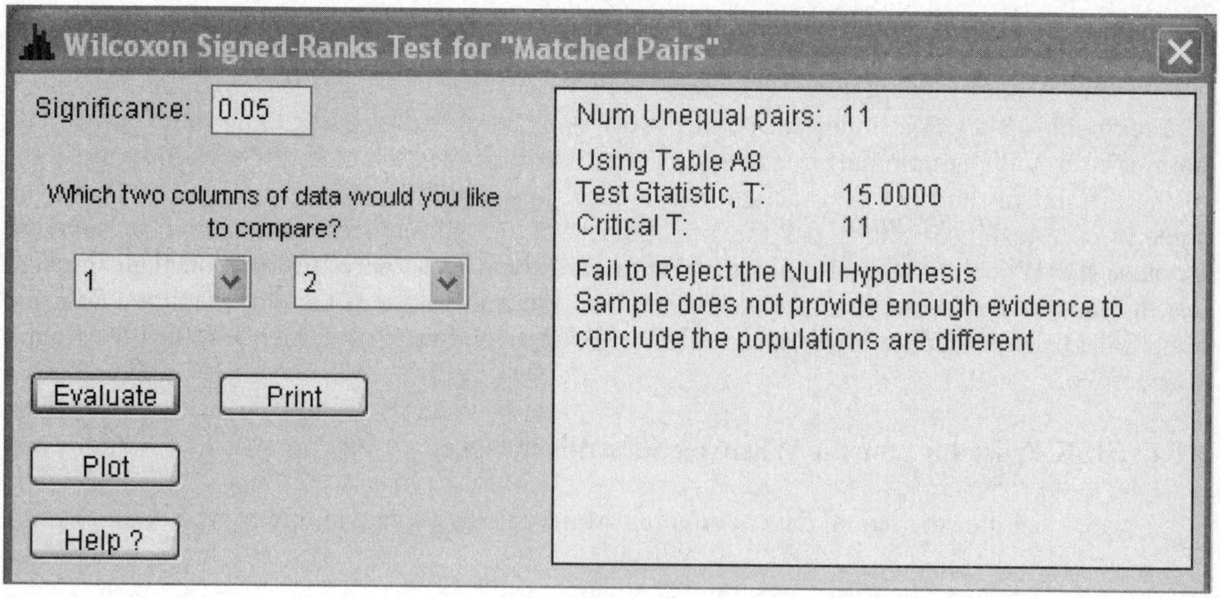

Wilcoxon Signed-Ranks Test for "Matched Pairs"

Significance: 0.05

Which two columns of data would you like to compare?

1 2

Evaluate Print

Plot

Help ?

Num Unequal pairs: 11

Using Table A8
Test Statistic, T: 15.0000
Critical T: 11

Fail to Reject the Null Hypothesis
Sample does not provide enough evidence to conclude the populations are different

13-4 Wilcoxon Rank-Sum Test

Section 13-4 of *Elementary Statistics* includes the following definition.

Definition
The **Wilcoxon rank-sum test** is a nonparametric test that uses ranks of sample data from two independent populations. It is used to test the null hypothesis that the two independent samples come from populations with equal medians. The alternative hypothesis is the claim that the two populations have different medians.

H_0: The two samples come from populations with equal medians.
H_1: The two samples come from populations with different medians.

First we describe the STATDISK procedure for conducting a Wilcoxon rank-sum test, then we illustrate it with an example.

STATDISK Procedure for the Wilcoxon Rank-Sum Test

1. Enter the two lists of values from the two independent samples in columns of the Statdisk data window.

2. Select **Analysis** from the main menu.

3. Select **Wilcoxon Tests** from the subdirectory.

4. You must now choose between the following two options.

   Signed-Ranks Test (Select this option with *matched pairs*.)
   Rank-Sum Test (Select this option with two *independent* samples.)

5. After selecting the **Rank-Sum Test** option, proceed to enter a significance level, such as 0.05 or 0.01.

6. Select the columns of the Statdisk data window that contain the two sets of independent sample data.

7. Click on the **Evaluate** button.

8. Click on **Plot** to display a graph that shows the test statistic and critical values.

Section 13-4 of the textbook includes two examples illustrating the Wilcoxon rank-sum test applied to BMI values for men and women. The first example uses a small part of the data listed in Data Set 1 in Appendix B, but the second example uses all of the BMI values given in Appendix B. The 40 BMI values for males and the 40 BMI values for females can be combined in the same STATDISK data window by using Copy/Paste with columns from the "Male Health" and "Female Health" data sets listed under "Health Exam Results." After using Copy/Paste to configure the data window so that the male BMI values are in column 1 and the female BMI values are in column 2, we use the Wilcoxon signed-ranks test to obtain the STATDISK results shown below. This display shows that the test statistic is $z = 1.03$, and the critical values are $z = \pm 1.96$. The test statistic is not in the critical region, so we fail to reject the null hypothesis. There does not appear to be a significant difference between the BMI values of males and females.

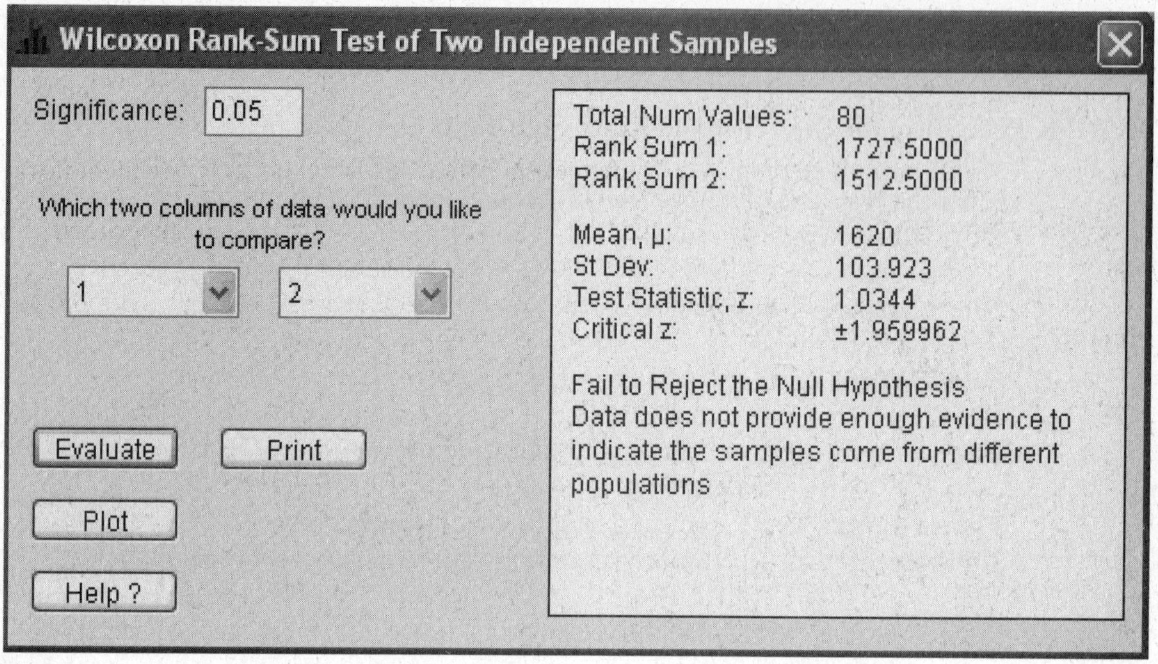

Wilcoxon Rank-Sum Test of Two Independent Samples

Significance: 0.05

Which two columns of data would you like to compare?

1 2

Evaluate Print

Plot

Help ?

Total Num Values:	80
Rank Sum 1:	1727.5000
Rank Sum 2:	1512.5000
Mean, µ:	1620
St Dev:	103.923
Test Statistic, z:	1.0344
Critical z:	±1.959962

Fail to Reject the Null Hypothesis
Data does not provide enough evidence to indicate the samples come from different populations

13-5 Kruskal-Wallis Test

Section 13-5 of *Elementary Statistics* includes this definition.

Definition
The **Kruskal-Wallis Test** (also called the ***H* test**) is a nonparametric test that uses ranks of sample data from three or more independent populations. It is used to test the null hypothesis that the independent samples come from populations with equal medians; the alternative hypothesis is the claim that the populations have medians that are not all equal.

H_0: The samples come from populations with equal medians.
H_1: The samples come from populations with medians that are not all equal.

We describe the STATDISK procedure for the Kruskal-Wallis test, then we illustrate it with an example.

STATDISK Procedure for the Kruskal-Wallis Test

1. Enter the samples of data in columns of the Statdisk data window.
2. Select **Analysis** from the main menu.
3. Select **Kruskal-Wallis test** from the submenu.
4. In the dialog box, enter a significance level, such as 0.05 or 0.01.
5. Select the columns containing the samples of data. Click on the boxes to insert or delete check marks. Columns with check marks are included in the calculations.
6. Click on the **Evaluate** button.
7. Click on **Plot** to display a graph that shows the test statistic and critical values.

The textbook illustrates the Kruskal-Wallis test with data obtained from an experiment with poplar trees. See the sample data in the table below. We will use the Kruskal-Wallis test to test the claim that the weights of poplar trees from the four treatment categories all have the same median. The STATDISK results are shown below the table of data.

Weights (kg) of Poplar Trees

Treatment			
None	Fertilizer	Irrigation	Fertilizer and Irrigation
0.15	1.34	0.23	2.03
0.02	0.14	0.04	0.27
0.16	0.02	0.34	0.92
0.37	0.08	0.16	1.07
0.22	0.08	0.05	2.38

```
Total Num Values:   20

Rank Sum 1:         45.0000
Rank Sum 2:         37.5000
Rank Sum 3:         42.5000
Rank Sum 4:         85.0000

Test Statistic, H:  8.2143
Critical H:         7.8147
P-value:            0.0418

Reject the Null Hypothesis
Data provides evidence that the samples come
from different populations
```

Important elements of the preceding display include the rank sums of 45, 37.5, 42.5, 85, the test statistic of $H = 8.2143$, the critical value of $H = 7.8147$, and the P-value of 0.0418. Remember that the Kruskal-Wallis test is a *right-tailed* test. Because the P-value of 0.0418 is less than the significance level of 0.05, we reject the null hypothesis. There is sufficient evidence to support a conclusion that the different treatments result in weights with medians that are not all equal.

13-6 Rank Correlation

Section 13-6 of *Elementary Statistics* introduces the use of *rank correlation*, which uses ranks in a procedure for determining whether there is some relationship between two variables.

Definition
The **rank correlation test** (or Spearman's rank correlation test) is a nonparametric test that uses ranks of sample data consisting of matched pairs. It is used to test for an association between two variables, so the null and alternative hypotheses are as follows (where ρ_s denotes the rank correlation coefficient for the entire population):

H_0: $\rho_s = 0$ (There is *no* correlation between the two variables.)
H_1: $\rho_s \neq 0$ (There is a correlation between the two variables.)

First we describe the STATDISK procedure, then we illustrate it with an example.

STATDISK Procedure for Rank Correlation

1. Enter the paired sample data in columns of the Statdisk data window.
2. Select **Analysis** from the main menu.
3. Select **Rank Correlation** from the submenu.
4. Enter a significance level, such as 0.05 or 0.01.
5. Select the columns containing the paired data to be used for the calculations.
6. Click on the **Evaluate** button.
7. Click on the **Plot** button to obtain a graph that shows the test statistic and critical values. The graph will be displayed only if the normal approximation is used (because $n > 30$).

Section 13-6 of the textbook includes paired data describing the ranks obtained when students and *U.S. News and World Report* magazine ranked colleges. Table 13–7 lists a sample of ranks. We want to determine whether there is a correlation between the rankings of students and the magazine. We will use STATDISK with a significance level of $\alpha = 0.05$.

The STATDISK display follows Table 13–7, and it shows that the test statistic is given by $r_s = 0.7143$. The critical values are shown as ±0.738. We therefore fail to reject the null hypothesis of no correlation. There is not sufficient evidence to support a claim of a correlation between the rankings of students and the magazine. Students and the magazine do not appear to agree on such rankings.

Table 13-7 Colleges Ranked by Students and *U.S. News and World Report*

College	Student Ranks	*U.S. News and World Report* Ranks
Harvard	1	1
Yale	2	2
Cal. Inst. of Tech.	3	5
M.I.T	4	4
Brown	5	7
Columbia	6	6
U. of Penn.	7	3
Notre Dame	8	8

```
Sample size, n:          8
Correlation Coeff, r:    0.7143
Critical r:              ±0.738

Fail to Reject the Null Hypothesis
Sample does not provide enough evidence
to conclude that the populations are rank
correlated
```

13-7 Runs Test for Randomness

Section 13-7 of *Elementary Statistics* includes these definitions.

Definitions
A **run** is a sequence of data having the same characteristic; the sequence is preceded and followed by data with a different characteristic or by no data at all.
The **runs test** uses the number of runs in a sequence of sample data to test for randomness in the order of the data.

First we describe the STATDISK procedure for the runs test for randomness, then we illustrate it with an example.

STATDISK Procedure for the Runs Test for Randomness

1. Using the original data, count the number of runs, the number of elements of the first type, and the number of elements of the second type.
2. Select **Analysis** from the main menu.
3. Select **Runs Test** from the submenu.
4. Make these entries in the dialog box:

   -Enter a significance level, such as 0.05 or 0.01.
   -Enter the number of runs.
   -Enter the number of elements of the first type.
   -Enter the number of elements of the second type.
5. Click on the **Evaluate** button.
6. Click on **Plot** to display a graph with the test statistic and critical values.

The textbook includes an example involving Boston rainfall amounts for Monday; those values are listed in Data Set 10 of Appendix B in the textbook. Letting D represent a dry day (with 0 rainfall) and letting R represent a rainy day (with a positive amount of rainfall), we get the sequence shown below. We want to determine whether rain on Mondays is random.

DDDDRDRDDRDDRDDDRDDRRRDDDDRDRDRRRDRDDDRDDDRDRDDRDDDR

The textbook describes the procedure for examining the above sequence to find these results:

G = number of runs = 30
n_1 = number of Ds = 33
n_2 = number of Rs = 19

Having found the number of elements of each type and the number of runs, we can use STATDISK to obtain the results shown below.

Runs Test for Randomness

Significance:	0.05
Num Runs:	30
Num Element 1:	33
Num Element 2:	19

Evaluate Print

Plot

Help ?

Num Runs, G: 30

Using Approximation
Mean, μ: 25.11538
St Dev: 3.306074
Test Statistic, z: 1.4775
Critical z: ±1.959962

Fail to Reject the Null Hypothesis
Not enough evidence to reject randomness

The above STATDISK display includes the value of u_G = 25.11538, σ_G = 3.306074, the test statistic of z = 1.4775, and the critical values of z= ±1.959962. These calculations are somewhat messy, as shown below.

$$\mu_G = \frac{2n_1n_2}{n_1+n_2} + 1 = \frac{2(33)(19)}{33+19} + 1 = 25.11538$$

$$\sigma_G = \sqrt{\frac{(2n_1n_2)(2n_1n_2 - n_1 - n_2)}{(n_1+n_2)^2(n_1+n_2-1)}}$$

$$= \sqrt{\frac{(2)(33)(19)[2(33)(19) - 33 - 19]}{(33+19)^2(33+19-1)}} = 3.306074$$

$$z = \frac{G - \mu_G}{\sigma_G} = \frac{30 - 25.115385}{3.306074} = 1.4775$$

STATDISK easily provides the results for these difficult calculations. Based on the results, we fail to reject the null hypothesis of randomness. It appears that on Mondays in Boston, the occurrence of rain is a random event. Big surprise there.

CHAPTER 13 EXPERIMENTS: Nonparametric Statistics

In Experiments 13–1 through 13–6, use STATDISK's sign test *program.*

13-1. ***Testing for a Difference Between Reported and Measured Male Heights*** As part of the National Health and Nutrition Examination Survey conducted by the Department of Health and Human Services, self–reported heights and measured heights were obtained for males aged 12–16. Listed below are sample results. Is there sufficient evidence to support the claim that there is a difference between self–reported heights and measured heights of males aged 12–16? Use a 0.05 significance level.

Reported height	68	71	63	70	71	60	65	64	54	63	66	72
Measured height	67.9	69.9	64.9	68.3	70.3	60.6	64.5	67.0	55.6	74.2	65.0	70.8

Test statistic:_____ Critical value:_____

Conclusion: ____

____

13–2. ***Testing for a Median Body Temperature of 98.6°F*** A pre–med student in a statistics class is required to do a class project. Intrigued by the body temperatures in Data Set 2 Appendix B, she plans to collect her own sample data to test the claim that the mean body temperature is less than 98.6°F, as is commonly believed. Because of time constraints, she finds that she has time to collect data from only 12 people. After carefully planning a procedure for obtaining a simple random sample of 12 healthy adults, she measures their body temperatures and obtains the results listed below. Use a 0.05 significance level to test the claim that these body temperatures come from a population with a median that is less than 98.6°F.

97.6 97.5 98.6 98.2 98.0 99.0 98.5 98.1 98.4 97.9 97.9 97.7

Test statistic:_____ Critical value:_____

Conclusion: ____

____

13–3. ***Testing for Median Underweight*** The Prince County Bottling Company supplies bottles of lemonade labeled 12 oz. When the Prince County Department of Weights and Measures tests a random sample of bottles, the amounts listed below are obtained. Using a 0.05 significance level, is there sufficient evidence to file a charge that the bottling company is cheating consumers by giving amounts with a median less than 12 oz?

11.4 11.8 11.7 11.0 11.9 11.9 11.5 12.0 12.1 11.9 10.9 11.3 11.5 11.5 11.6

Test statistic:_____ Critical value:_____

Conclusion: ______________________________

____

13–4. ***Nominal Data: Survey of Voters*** In a survey of 1002 people, 701 said that they voted in the recent presidential election (based on data from ICR Research Group). Is there sufficient evidence to support the claim that the majority of people say that they voted in the election?

Test statistic:_____ Critical value:_____

Conclusion: ______________________________

____

13–5. ***Nominal Data: Smoking and Nicotine Patches*** In one study of 71 smokers who tried to quit smoking with nicotine patch therapy, 41 were smoking one year after the treatment (based on data from "High-Dose Nicotine Patch Therapy," by Dale et al., *Journal of the American Medical Association,* Vol. 274, No. 17). Use a 0.05 significance level to test the claim that among smokers who try to quit with nicotine patch therapy, the majority are smoking a year after the treatment.

Test statistic:_____ Critical value:_____

Conclusion: ______________________________

____

13–6. ***Testing for Difference Between Forecast and Actual Temperatures*** Refer to Data Set 8 in Appendix B and use the actual high temperatures and the three day forecast high temperatures. Does there appear to be a difference?

Test statistic:_____ Critical value:_____

Conclusion: ______________________________

____

13–7. ***Using Parametric Test*** Repeat Experiment 13–6 using an appropriate *parametric* test from Chapter 9. Compare the results from the parametric test and the sign test. Do the results lead to the same conclusion? Is either test more sensitive to the differences between pairs of data?

____

____

13–8. ***Sign Test vs. Wilcoxon Signed–Ranks Test*** Repeat Experiment 13-6 by using the Wilcoxon signed-ranks test for matched pairs. Enter the STATDISK results below, and compare them to the sign test results obtained in Experiment 13-6. Specifically, how do the results reflect the fact that the Wilcoxon signed-ranks test uses more information?

Test statistic:_____ Critical value:_____

Conclusion: ______________________________

____

Comparison: ______________________________

____

13–9. ***Testing for Drug Effectiveness*** Captopril is a drug designed to lower systolic blood pressure. When subjects were tested with this drug, their systolic blood pressure readings (in mm of mercury) were measured before and after the drug was taken, with the results given in the accompanying table (based on data from "Essential Hypertension: Effect of an Oral Inhibitor of Angiotensin-Converting Enzyme," by MacGregor et al., *British Medical Journal,* Vol. 2). Is there sufficient evidence to support the claim that the drug has an effect? Does Captopril appear to lower systolic blood pressure? Use the Wilcoxon signed–ranks test.

Subject	A	B	C	D	E	F	G	H	I	J	K	L
Before	200	174	198	170	179	182	193	209	185	155	169	210
After	191	170	177	167	159	151	176	183	159	145	146	177

Test statistic:_____ Critical value:_____

Conclusion: ______________________________

____

13–10. ***Are Severe Psychiatric Disorders Related to Biological Factors?*** One study used X-ray computed tomography (CT) to collect data on brain volumes for a group of patients with obsessive-compulsive disorders and a control group of healthy persons. The accompanying list shows sample results (in milliliters) for volumes of the right cordate (based on data from "Neuroanatomical Abnormalities in Obsessive-Compulsive Disorder Detected with Quantitative X-Ray Computed Tomography," by Luxenberg et al., *American Journal of Psychiatry,* Vol. 145, No. 9). Use the Wilcoxon rank–sum test with a 0.01 significance level to test the claim that obsessive-compulsive patients and healthy persons have the same brain volumes. Based on this result, can we conclude that obsessive-compulsive disorders have a biological basis?

Obsessive-compulsive patients				**Control group**			
0.308	0.210	0.304	0.344	0.519	0.476	0.413	0.429
0.407	0.455	0.287	0.288	0.501	0.402	0.349	0.594
0.463	0.334	0.340	0.305	0.334	0.483	0.460	0.445

Test statistic:_____ Critical value:_____

Conclusion: ____

____

13–11. ***Testing the Anchoring Effect*** Randomly selected statistics students were given five seconds to estimate the value of a product of numbers with the results given in the accompanying table. (See the Cooperative Group Activities at the end of Chapter 2.) Is there sufficient evidence to support the claim that the two samples come from populations with different distributions? Use the Wilcoxon rank–sum test.

Estimates from Students Given $1 \times 2 \times 3 \times 4 \times 5 \times 6 \times 7 \times 8$

1560 169 5635 25 842 40,320 5000 500 1110 10,000
200 1252 4000 2040 175 856 42,200 49,654 560 800

Estimates from Students Given $8 \times 7 \times 6 \times 5 \times 4 \times 3 \times 2 \times 1$

100,000 2000 42,000 1500 52,836 2050 428 372 300 225
64,582 23,410 500 1200 400 49,000 4000 1876 3600 354 750
640

Test statistic:_____ Critical value:_____

Conclusion: ____

____

13–12. ***Does Weight of a Car Affect Head Injuries in a Crash?*** Data were obtained from car crash experiments conducted by the National Transportation Safety Administration. New cars were purchased and crashed into a fixed barrier at 35 mi/h, and measurements were recorded for the dummy in the driver's seat. Use the sample data listed below to test for differences in head injury measurements (in hic) among the four weight categories. Is there sufficient evidence to conclude that head injury measurements for the four car weight categories are not all the same? Do the data suggest that heavier cars are safer in a crash?

Subcompact:	681	428	917	898	420
Compact:	643	655	442	514	525
Midsize:	469	727	525	454	259
Full–Size:	384	656	602	687	360

Test statistic:_____ Critical value:_____

Conclusion: ____

____

13–13. ***Do All Colors of M&Ms Weigh the Same?*** Refer to Data Set 13 in Appendix B. At the 0.05 significance level, test the claim that the weights of M&Ms are the same for each of the six different color populations. If it is the intent of Mars, Inc., to make the candies so that the different color populations are the same, do your results suggest that the company has a problem that requires corrective action?

Test statistic:_____ Critical value:_____

Conclusion: ____

____

12–14. ***Homerun Distances*** Refer to Data Set 17 in Appendix B. Consider the homerun distances to be samples randomly selected from populations. Use a 0.05 significance level to test the claim that the populations of distances of homeruns hit by Barry Bonds, Mark McGwire, and Sammy Sosa have the same median.

Test statistic:_____ Critical value:_____

Conclusion: ____

____

13–15. ***Business School Rankings*** *Business Week* magazine ranked business schools two different ways. Corporate rankings were based on surveys of corporate recruiters, and graduate rankings were based on surveys of MBA graduates. The table below is based on the results for 10 schools. Is there a correlation between the corporate rankings and the graduate rankings? Use a significance level of $\alpha = 0.05$.

School	PA	NW	Chi	Sfd	Hvd	MI	IN	Clb	UCLA	MIT
Corporate ranking	1	2	4	5	3	6	8	7	10	9
Graduate ranking	3	5	4	1	10	7	6	8	2	9

Test statistic:_____ Critical value:_____

Conclusion: ____________________

____

13–16. ***Correlation Between Restaurant Bills and Tips*** Students of the author collected sample data consisting of amounts of restaurant bills and the corresponding tip amounts. The data are listed below. Use rank correlation to determine whether there is a correlation between the amount of the bill and the amount of the tip.

Bill (dollars)	33.46	50.68	87.92	98.84	63.60	107.34
Tip (dollars)	5.50	5.00	8.08	17.00	12.00	16.00

Test statistic:_____ Critical value:_____

Conclusion: ____________________

____

13–17. ***Supermodel Heights and Weights*** Listed below are heights (in inches) and weights (in pounds) for supermodels Michelle Alves, Nadia Avermann, Paris Hilton, Kelly Dyer, Christy Turlington, Bridget Hall, Naomi Campbell, Valerie Mazza, and Kristy Hume. Is there a correlation between height and weight? If there is a correlation, does it mean that there is a correlation between height and weight of all adult women?

Height (in.)	70	70.5	68	65	70	70	70	70	71
Weight (lb)	117	119	105	115	119	127	113	123	115

Test statistic:_____ Critical value:_____

Conclusion: ____________________

____

13–18. ***Buying a TV Audience*** The *New York Post* published the annual salaries (in millions) and the number of viewers (in millions), with results given below for Oprah Winfrey, David Letterman, Jay Leno, Kelsey Grammer, Barbara Walters, Dan Rather, James Gandolfini, and Susan Lucci, repsectively. Is there a correlation?

Salary	100	14	14	35.2	12	7	5	1
Viewers	7	4.4	5.9	1.6	10.4	9.6	8.9	4.2

Test statistic:_____ Critical value:_____

Conclusion: ______________________________

____

13–19. ***Home Sales*** Refer to Data Set 18 in Appendix B and test for a correlation between the list price and selling price of homes.

Test statistic:_____ Critical value:_____

Conclusion: ______________________________

____

13–20. ***Cholesterol and Body Mass Index*** Refer to Data Set 1 in Appendix B and use the cholesterol levels and body mass index values of the 40 women. Is there a correlation between cholesterol level and body mass index?

Test statistic:_____ Critical value:_____

Conclusion: ______________________________

____

13–21. ***Testing for Randomness of Survey Respondents*** When selecting subjects to be surveyed about Blue Fang's *Zoo Tycoon* game, the subjects were selected in a sequence with the genders listed below. Does it appear that the subjects were randomly selected according to gender?

M M F F F M F M M M M F F M M F F F F M F

Test statistic:_____ Critical value:_____

Conclusion: ______________________________

____

13–22. ***Testing for Randomness in Dating Prospects*** Fred has had difficulty getting dates with women, so he is abandoning his strategy of careful selection and replacing it with a desperate strategy of random selection. In pursuing dates with randomly selected women, Fred finds that some of them are unavailable because they are married. Fred, who has an abundance of time for such activities, records and analyzes his observations. Given the results listed below (where M denotes married and S denotes single), what should Fred conclude about the randomness of the women he selects?

M M M M S S S S S S M M M M M S S S M M M M M M M M M M S S S S

Test statistic:_____ Critical value:_____

Conclusion: ____

____

13–23. ***Testing for Randomness of Baseball World Series Victories*** Test the claim that the sequence of World Series wins by American League and National League teams is random. Given below are recent results, with American and National League teams represented by A and N, respectively.

A N A N N N A A A A N A A A A N A N N A A N N A A A A N A N
N A A A A A N A N A N A N A A A A A A A A N N A N A N N A A N
N N A N A N A N A A A N N A A N N N N A A A N A N A N A A A
N A N A A A N A N A

Test statistic:_____ Critical value:_____

Conclusion: ____

____

13–24. ***Testing for Randomness of Presidential Election Winners*** For a recent sequence of presidential elections, the political party of the winner is indicated by D for Democrat and R for Republican. Does it appear that we elect Democrat and Republican candidates in a sequence that is random?

R R D R D R R R R D D R R R D D D
D D R R D D R R D R R R D D R R

Test statistic:_____ Critical value:_____

Conclusion: ____

____

13–25. ***Activities with STATDISK: Rank Correlation***

After a successful bowling tournament, six couples dine at a restaurant. Each couple leaves a 15% tip. A week later the same six couples dine at the same restaurant, order the same meals, and have the same waiter. Being in a jovial mood and wanting to see the waiter's reaction, they decide to scramble their tips. We will explore the effects of each of these two events (regular tips and scrambled tips) on rank correlation.

Enter the following data in columns 1 through 3 of the STATDISK data window, and observe the effect on the Spearman rank correlation coefficient. Use a 0.05 significance level.

Bill	Regular Tips (First visit)	Scrambled Tips (Second visit)
30.90	4.75	7.00
55.89	8.00	10.00
101.15	15.00	12.00
78.80	12.00	15.00
67.50	10.00	8.00
46.25	7.00	4.75

First visit: Regular Tips

Using the bill amounts and the tip amounts from the first visit, find the rank correlation coefficient r_s: ____________________

What do you conclude? ____

Second visit: Scrambled Tips

Using the bill amounts and the tip amounts from the second visit, find the rank correlation coefficient r_s: ____________________

What do you conclude? ____

What do the results from the two events indicate about rank correlation?

____

Suppose that one couple feels exceptionally generous and changes their tip from $4.75 to $500. That tip of $500 clearly becomes an outlier. After replacing both values of $4.75 in the above table by values of $500, repeat the analyses and enter the results below.

(continued)

First visit: Regular Tips

Using the bill amounts and the tip amounts from the first visit, find the rank correlation coefficient r_s: ______________________

What do you conclude? ____

Second visit: Scrambled Tips

Using the bill amounts and the tip amounts from the second visit, find the rank correlation coefficient r_s: ______________________

What do you conclude? ____

What do these new results indicate about the effect that an outlier has on rank correlation?

____

____

Statistical Process Control

The major topics of Chapter 14 from *Elementary Statistics* are run charts, control charts for variation, control charts for mean, and control charts for attributes. Although STATDISK is not programmed to generate run charts or control charts, there are ways to obtain them by using STATDISK's scatterplot feature. There are other software packages, such as Minitab, that are easier to use when constructing run charts or control charts, but this chapter focuses on the use of STATDISK.

14-1 Run Charts

In Chapter 14 from *Elementary Statistics* we define **process data** to be data arranged according to some time sequence, such as the data in Table 14-1, which is reproduced below. At the Altigauge Manufacturing Company, four altimeters are randomly selected from production on each of 20 consecutive business days, and Table 14–1 lists the *errors* (in feet) when they are tested in a pressure chamber that simulates an altitude of 1000 ft. The error measurements are displayed in Table 14–1 with bold text. On day 1, for example, the actual altitude readings for the four selected altimeters are 1002 ft, 992 ft, 1005 ft, and 1011 ft, so the corresponding errors (in feet) are 2, –8, 5, and 11. Those four errors have a mean of 2.50 ft, a median of 3.5 ft, a range of 19 ft, and a standard deviation of 7.94 ft.

Table 14–1 Aircraft Altimeter Errors (in feet)

Day					Mean	Median	Range	St. Dev.
1	**2**	**-8**	**5**	**11**	2.50	3.5	19	7.94
2	**-5**	**2**	**6**	**8**	2.75	4.0	13	5.74
3	**6**	**7**	**-1**	**-8**	1.00	2.5	15	6.98
4	**-5**	**5**	**-5**	**6**	0.25	0.0	11	6.08
5	**9**	**3**	**-2**	**-2**	2.00	0.5	11	5.23
6	**16**	**-10**	**-1**	**-8**	-0.75	-4.5	26	11.81
7	**13**	**-8**	**-7**	**2**	0.00	-2.5	21	9.76
8	**-5**	**-4**	**2**	**8**	0.25	-1.0	13	6.02
9	**7**	**13**	**-2**	**-13**	1.25	2.5	26	11.32
10	**15**	**7**	**19**	**1**	10.50	11.0	18	8.06
11	**12**	**12**	**10**	**9**	10.75	11.0	3	1.50
12	**11**	**9**	**11**	**20**	12.75	11.0	11	4.92
13	**18**	**15**	**23**	**28**	21.00	20.5	13	5.72
14	**6**	**32**	**4**	**10**	13.00	8.0	28	12.91
15	**16**	**-13**	**-9**	**19**	3.25	3.5	32	16.58
16	**8**	**17**	**0**	**13**	9.50	10.5	17	7.33
17	**13**	**3**	**6**	**13**	8.75	9.5	10	5.06
18	**38**	**-5**	**-5**	**5**	8.25	0.0	43	20.39
19	**18**	**12**	**25**	**-6**	12.25	15.0	31	13.28
20	**-27**	**23**	**7**	**36**	9.75	15.0	63	27.22

To use STATDISK for generating a run chart, pair the data with the consecutive positive integers, then generate a scatter diagram (available in the Data menu). The table below shows the first two rows of data values paired with 1, 2, 3, . . . , 8.

x	1	2	3	4	5	6	7	8
y	2	−8	5	11	−5	2	6	8

This table includes only the first eight values in Table 14-1, but it can be easily extended to include all of the data. Continue with this procedure to enter the consecutive table values matched with the positive integers. Enter the paired data in columns of the Statdisk Data Window.

Shown below is the scatterplot of the 80 sample values from Table 14-1 paired with the positive integers 1, 2, 3, . . . , 80. The scatterplot was obtained by selecting **Data**, then **Scatterplot**, and the points were plotted *without* the regression line included.

STATDISK Run Chart

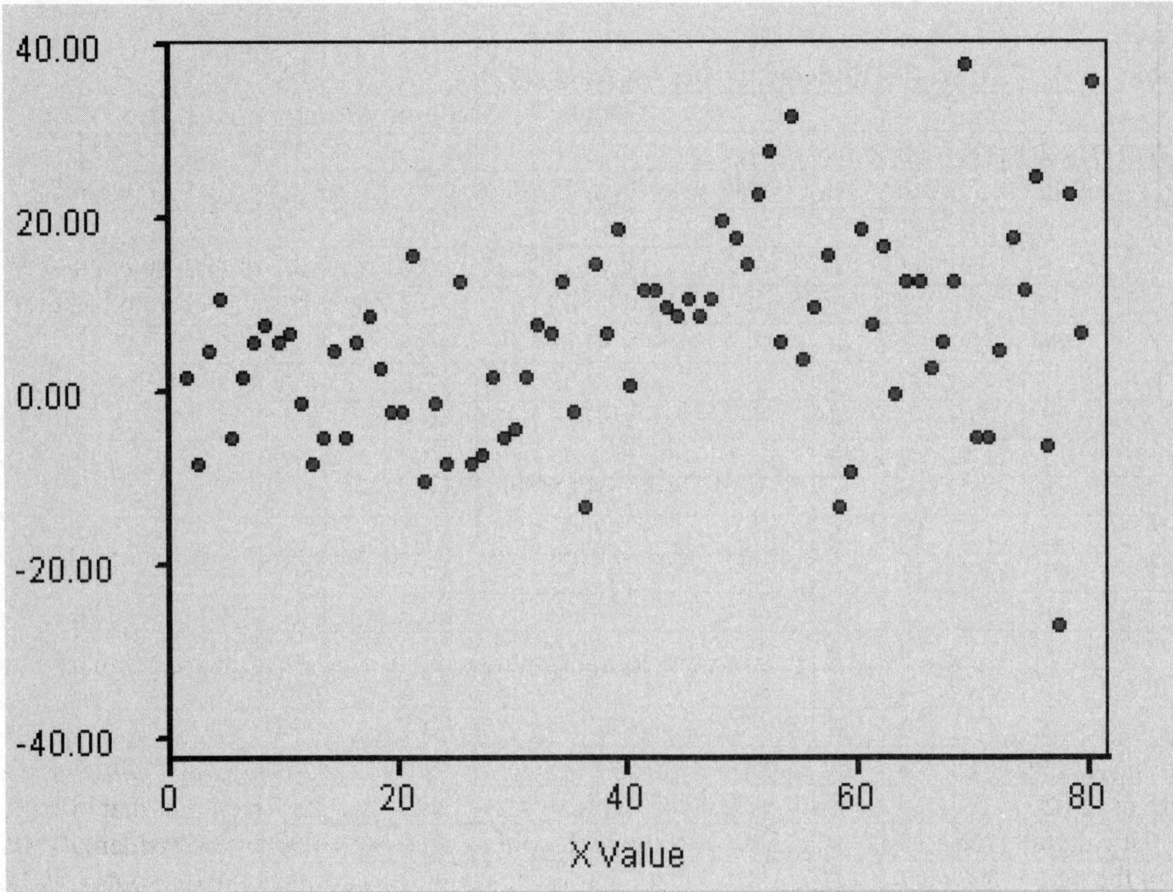

STATDISK does not connect the points as shown in the above display, but they can be easily connected to get a run chart such as the Minitab–generated run chart shown below. Go ahead and connect the dots in the above STATDISK display; it's more fun than human beings should be allowed to have. In so doing, you will effectively create the same run chart included in Section 14-2 of the textbook.

Minitab–Generated Run Chart

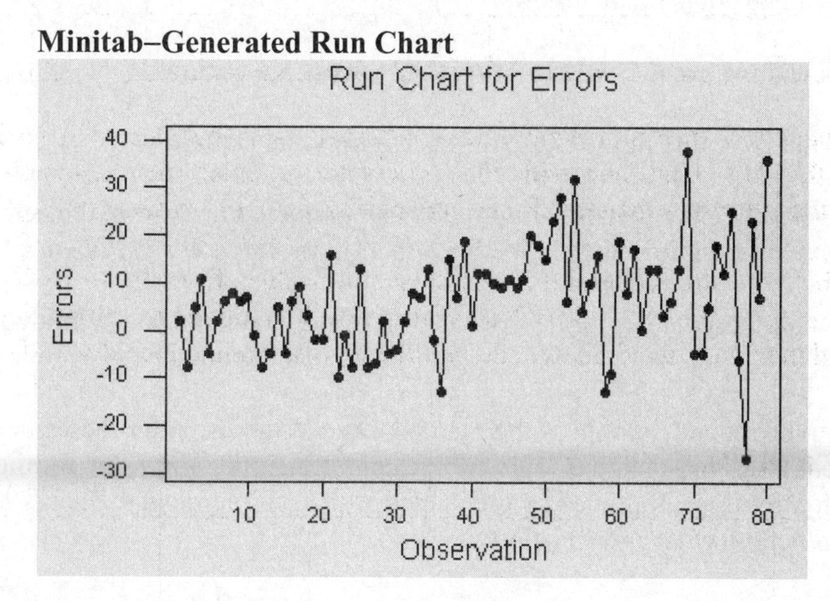

STATDISK Procedure for Run Chart

Based on the preceding example, here are the steps for using STATDISK to create a run chart:

1. Enter the process data in a column of the Statdisk Data Window. Also enter the consecutive positive integers 1, 2, 3, . . . (stopping when the column has the same length as the column of the process data).

2. Select **Analysis** from the main menu.

3. Select the menu item of **Data**, then select **Scatterplot**.

4. Click on the box next to "Regression Line" so that there is no check mark in that box, then click on the **Plot** button.

5. Connect the points in order from left to right, as shown in the above Minitab run chart.

To *interpret* the run chart, use the criteria described in Section 14-2 of *Elementary Statistics*. See Figure 14-2 in *Elementary Statistics* for typical samples of process patterns that are *not* statistically stable. For the preceding run chart, the points at the left fluctuate considerably less than the points farther to the right. It appears that the manufacturing process started out well, but deteriorated as time passed.

14-2 Control Charts for Variation

Section 14-2 of the textbook describes ***R* charts** (sequential plots of ranges) for monitoring *variation* in a process. Using the data of Table 14-1, for example, the *R* chart is a plot of the ranges 78, 77, . . . , 31. Again, STATDISK does not have a feature designed specifically for *R* charts, but they can be obtained as follows.

STATDISK Procedure for *R* Chart to Monitor Process Variation

1. Find the value of the range for each individual sample.

2. Determine the values to be used for the upper control limit, the centerline, and the lower control limit. (See Section 14-2 in *Elementary Statistics*.)

3. Enter these values in the first column of the Statdisk Data Window: 0, 0, 0, 1, 2, 3, . . . (stopping when you reach the number of sample ranges).

4. In the second column of the Statdisk Data Window, enter the value of the upper control limit, then the value for the centerline, then the value for the lower control limit, followed by the values of the sample ranges. That is, follow the general format shown below.

Column 1	Column 2
0	(Value of upper control limit)
0	(Value of $\overline{R}$ for the centerline)
0	(Value of lower control limit)
1	(Value of first sample range)
2	(Value of second sample range)
3	(Value of third sample range)
and so on	

Section 14–2 of *Elementary Statistics* shows how to determine the values of the upper control limit (48.4), the centerline (21.2), and the lower control limit (0.0). Use those three values followed by the list of sample ranges (19, 13, 15, . . . , 63) as shown in the following list and the Statdisk data window on the next page.

Column 1	Column 2	
0	48.4	←Upper control limit
0	21.2	←Centerline
0	0.0	←Lower control limit
1	19	←First sample range
2	13	←Second sample range
3	15	←Third sample range
and so on		

Row	1	2
1	0	48.4
2	0	21.2
3	0	0.0
4	1	19
5	2	13
6	3	15
7	4	11
8	5	11
9	6	26
10	7	21
11	8	13
12	9	26
13	10	18
14	11	3
15	12	11
16	13	13

5. Click on **Data**, then select **Scatterplot**.

6. Click on the "Regression Line" box so that there is no check mark there, then click on **Plot**.

7. The three leftmost points will be stacked above 0. Use those points to position the upper control limit, the centerline, and the lower control limit. For the remaining points beginning above 1, connect the points in order from left to right.

The STATDISK *R* chart for the data in Table 14-1 is shown below. The graph was modified to include lines and labels for the upper control limit, the centerline, and the lower control limit. Note that the three thick lines were located by including these three points as the first three points in the list of paired data: (0, 48.4), (0, 21.2), (0, 0.00).

We can interpret the *R* chart by applying the three out-of-control criteria given in the textbook:

1. There is no pattern, trend, or cycle that is obviously not random.
2. No point lies beyond the upper or lower control limits.
3. There are not 8 consecutive points all above or all below the center line.

The following STATDISK *R* chart leads to the conclusion that the process is out of statistical control because there is a point beyond the upper control limit.

STATDISK *R* Chart

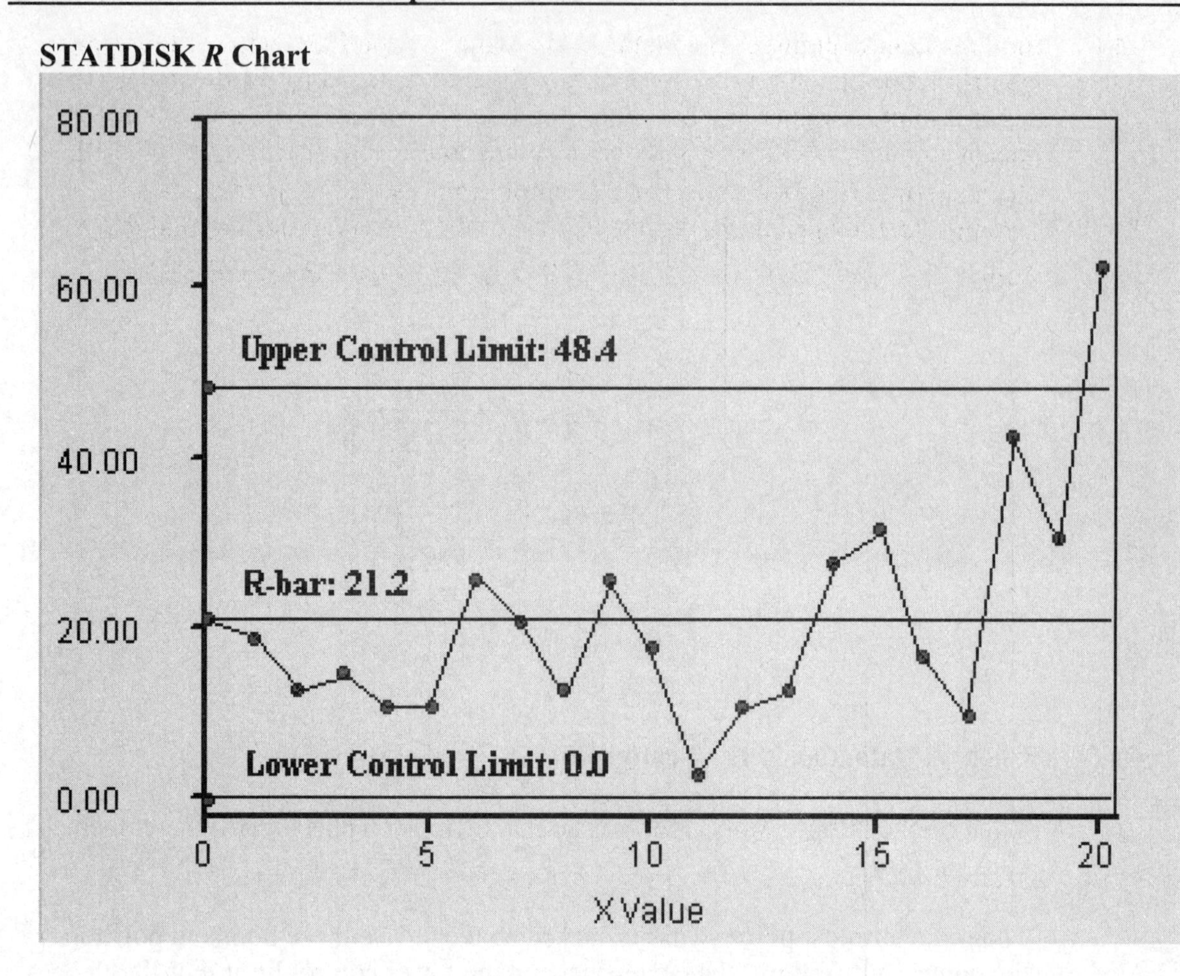

14-3 Control Charts for Mean

Control charts for $\bar{x}$ can be constructed by using the same methods described for *R* charts. You must first find the sample means and the values to be used for the upper control limit, the centerline, and the lower control limit. See Section 14-2 in *Elementary Statistics* for the procedures that can be used to find these values.

STATDISK Procedure for $\bar{x}$ Chart to Monitor Process Mean

1. Find the value of the mean for each individual sample.

2. Determine the values to be used for the upper control limit, the centerline, and the lower control limit. (See Section 14-2 in *Elementary Statistics*.)

3. Enter these values in the first column of the Statdisk data window:
   0, 0, 0, 1, 2, 3, . . . (stopping when reaching the number of sample means).

4. In the second column of the Statdisk data window, enter the value of the upper control limit, then the value for the centerline, then the value for the lower control limit, followed by the values of the sample means. Using the sample means in Table 14-1 along with the locations of the upper control limit (21.9), the centerline (6.45), and the lower control limit (–9.0), the Statdisk data window should appear as shown below. (For the procedure used to find these values of 21.9, 6.45, and –9.0, see Section 14-2 of *Elementary Statistics*.)

Row	1	2
1	0	21.9
2	0	6.45
3	0	-9.0
4	1	2.50
5	2	2.75
6	3	1.00
7	4	0.25
8	5	2.00
9	6	-0.75
10	7	0.00
11	8	0.25
12	9	1.25
13	10	10.50
14	11	10.75
15	12	12.75
16	13	21.00

5. Click on **Data**, then select **Scatterplot**.

6. Click on the "Regression Line" box so that there is no check mark, then click on **Plot**.

7. The three leftmost points will be stacked above 0. Use those points to position the upper control limit, the centerline, and the lower control limit. For the remaining points beginning above 1, connect the points in order from left to right.

Shown on the next page is the STATDISK $\overline{x}$ chart for the data in Table 14-1. The graph was modified to include lines and labels for the upper control limit, the centerline, and the lower control limit. Note that the three thick lines were located by including these three points as the first three points in the list of paired data: (0, 21.9), (0, 6.45), (0, –9.0).

We can interpret the $\overline{x}$ chart by applying the three out-of-control criteria given in the textbook. We conclude that the mean in this process is out of statistical control because there does appear to be a pattern of an upward trend, and there 8 consecutive points all below the center line.

STATDISK $\overline{x}$ Chart

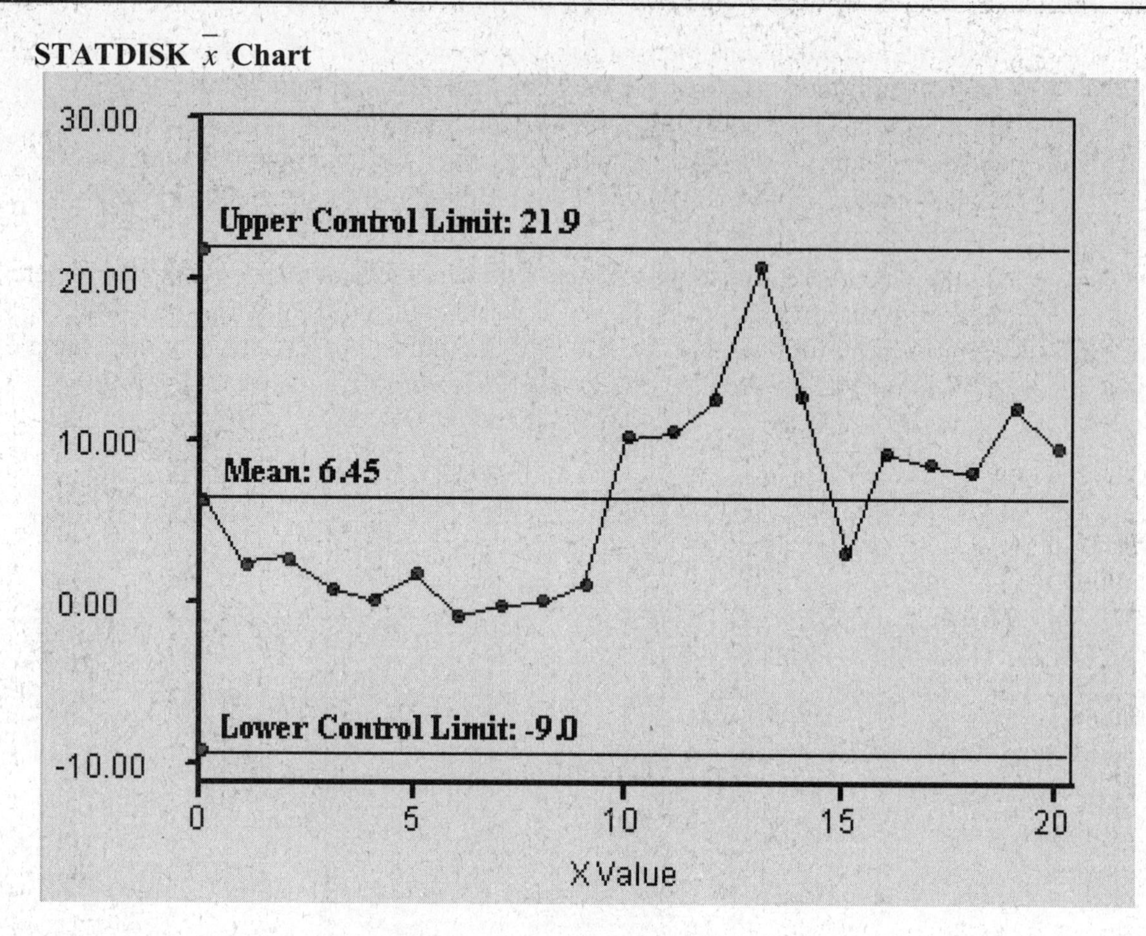

14-4 Control Charts for Attributes

A control chart for attributes (or p chart) can also be constructed by using the same procedure for R charts and $\overline{x}$ charts. A p chart is very useful in monitoring some process proportion, such as the proportions of defects over time. The example in Section 14-3 of the textbook involves the numbers of defective altimeters in successive batches of 100 each. The numbers of defects are listed below.

Defects: 2 0 1 3 1 2 2 4 3 5 3 7

Refer to the example in Section 14-3 of the textbook and use the following procedure:

STATDISK Procedure for *p* Charts

1. Collect the list of sample proportions. For the above sample data, the proportions are 0.02, 0, 0.01, 0.03, 0.01, 0.02, 0.02, 0.04, 0.03, 0.05, 0.03, and 0.07.

2. Determine the values to be used for the upper control limit, the centerline, and the lower control limit. (See Section 14-3 in *Elementary Statistics*.) For the above sample data, the upper control limit is at 0.0766, the centerline is at 0.0275, and

the lower control limit is at 0.

3. Enter these values in the first column of the Statdisk data window: 0, 0, 0, 1, 2, 3, ... (stop when reaching the number of sample proportions).

4. In the second column of the Statdisk data window, enter the value of the upper control limit, then the value for the centerline, then the value for the lower control limit, followed by the values of the sample proportions. For the example in Section 14-3 of the textbook we should make entries in the dialog box as shown below.

Row	1	2
1	0	0.0766
2	0	0.0275
3	0	0
4	1	0.02
5	2	0
6	3	0.01
7	4	0.03
8	5	0.01
9	6	0.02
10	7	0.02
11	8	0.04
12	9	0.03
13	10	0.05
14	11	0.03
15	12	0.07

5. Click on **Data**, then select **Scatterplot**.

6. Click on the "Regression Line" box to remove the check mark, then click on **Plot**.

7. The three leftmost points will be stacked above 0. Use those points to position the upper control limit, the centerline, and the lower control limit. For the remaining points beginning above 1, connect the points in order from the left.

The following control chart for p can be interpreted by using the same three out-of-control criteria listed in Section 14-2 of the textbook. Using those criteria, we conclude that this process is out of statistical control because there appears to be an upward trend. The company should take immediate action to correct the increasing proportion of defects.

STATDISK *p* Chart

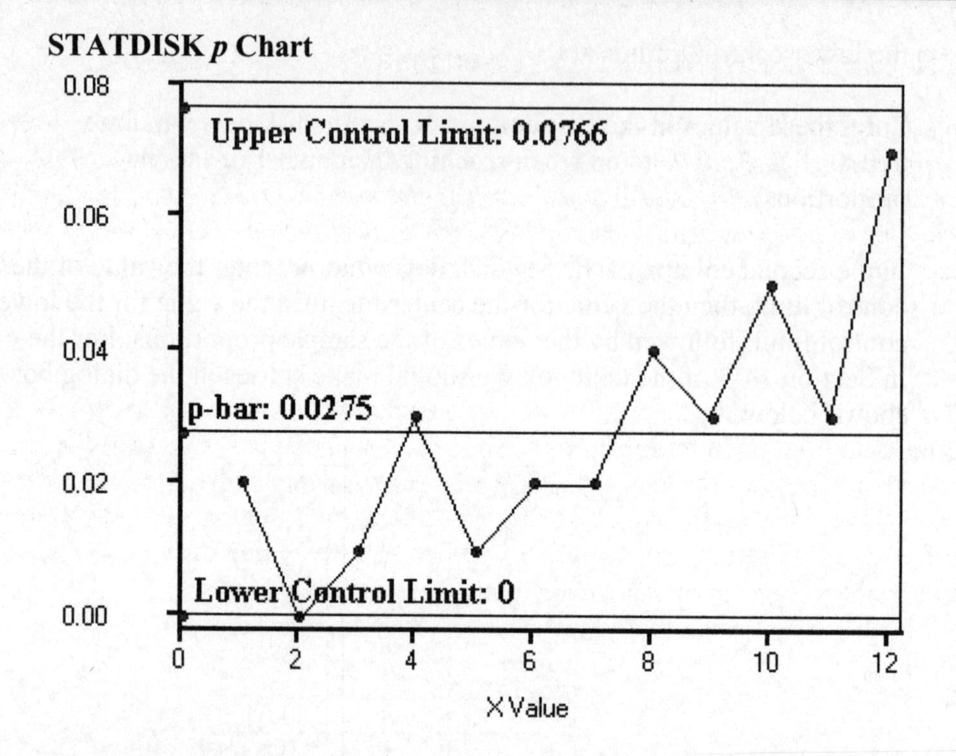

Minitab and some other statistical software automatically generate run charts and control charts that are so important for monitoring process data over time. Shown below is the Minitab *p* chart that is automatically generated with lines for the upper control limit, centerline, and lower control limit. The use of such charts is increasing as more businesses recognize that this statistical tool can be effective in increasing quality and lowering costs.

Minitab *p* Chart

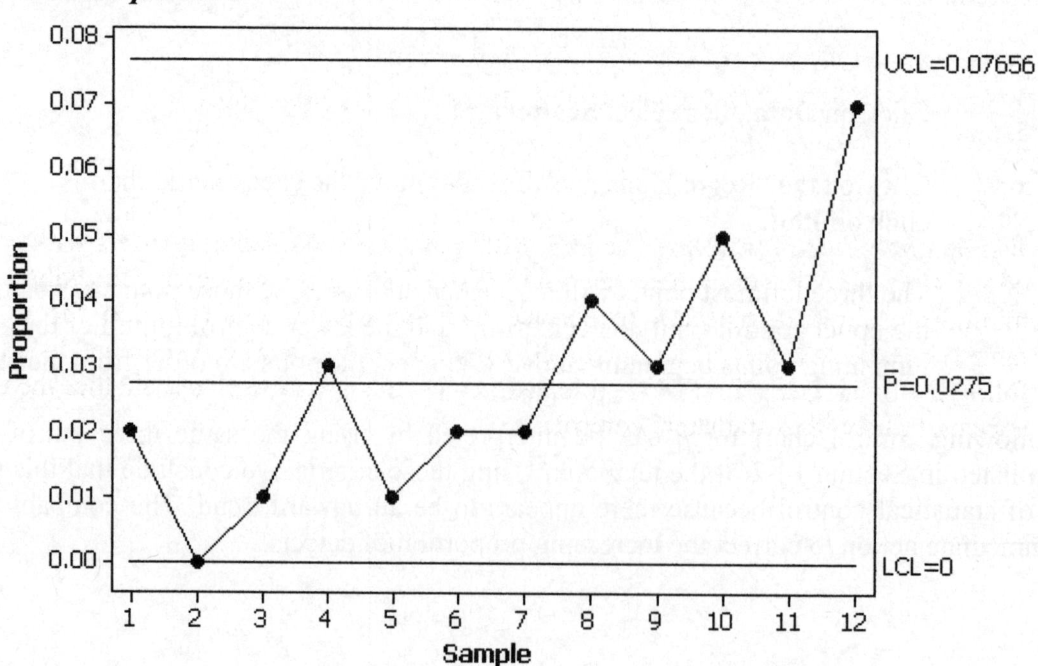

CHAPTER 14 EXPERIMENTS: Statistical Process Control

Constructing Control Charts for Aluminum Cans *Experiments 1 and 2 are based on the axial loads(in pounds) of aluminum cans that are 0.0109 in. thick, as listed in Data Set 15 in Appendix B of the textbook. An axial load of a can is the maximum weight supported by its side, and it is important to have an axial load high enough so that the can isn't crushed when the top lid is pressed into place. The data are from a real manufacturing process, and they were provided by a student who used an earlier version of this book.*

14–1. ***R Chart*** On each day of production, seven aluminum cans with thickness 0.0109 in. were randomly selected and the axial loads were measured. The ranges for the different days are listed below, but they can also be found from the values given in Data Set 15 in Appendix B of the textbook. Construct an R chart and determine whether the process variation is within statistical control. If it is not, identify which of the three out-of-control criteria lead to rejection of statistically stable variation.

78 77 31 50 33 38 84 21 38 77 26 78 78
17 83 66 72 79 61 74 64 51 26 41 31

14–2. $\bar{x}$ ***Chart*** On each day of production, seven aluminum cans with thickness 0.0109 in. were randomly selected and the axial loads were measured. The means for the different days are listed below, but they can also be found from the values given in Data Set 15 in Appendix B of the textbook. Construct an $\bar{x}$ chart and determine whether the process mean is within statistical control. If it is not, identify which of the three out-of-control criteria lead to rejection of statistically stable variation.

252.7 247.9 270.3 267.0 281.6 269.9 257.7 272.9 273.7 259.1
275.6 262.4 256.0 277.6 264.3 260.1 254.7 278.1 259.7 269.4
266.6 270.9 281.0 271.4 277.3

14-3. ***Weights of Minted Quarters*** The U.S. Mint has a goal of making quarters with a weight of 5.670 g, but any weight between 5.443 g and 5.897 g is considered acceptable. A new minting machine is placed into service and the weights are recorded for a quarter randomly selected every 12 min for 20 consecutive hours. The results are listed in the following table. Use STATDISK to construct a run chart. Determine whether the process appears to be within statistical control.

Hour	Weight of Quarter (grams)					Mean	Range
1	5.639	5.636	5.679	5.637	5.691	5.6564	0.055
2	5.655	5.641	5.626	5.668	5.679	5.6538	0.053
3	5.682	5.704	5.725	5.661	5.721	5.6986	0.064
4	5.675	5.648	5.622	5.669	5.585	5.6398	0.090
5	5.690	5.636	5.715	5.694	5.709	5.6888	0.079
6	5.641	5.571	5.600	5.665	5.676	5.6306	0.105
7	5.503	5.601	5.706	5.624	5.620	5.6108	0.203
8	5.669	5.589	5.606	5.685	5.556	5.6210	0.129
9	5.668	5.749	5.762	5.778	5.672	5.7258	0.110
10	5.693	5.690	5.666	5.563	5.668	5.6560	0.130
11	5.449	5.464	5.732	5.619	5.673	5.5874	0.283
12	5.763	5.704	5.656	5.778	5.703	5.7208	0.122
13	5.679	5.810	5.608	5.635	5.577	5.6618	0.233
14	5.389	5.916	5.985	5.580	5.935	5.7610	0.596
15	5.747	6.188	5.615	5.622	5.510	5.7364	0.678
16	5.768	5.153	5.528	5.700	6.131	5.6560	0.978
17	5.688	5.481	6.058	5.940	5.059	5.6452	0.999
18	6.065	6.282	6.097	5.948	5.624	6.0032	0.658
19	5.463	5.876	5.905	5.801	5.847	5.7784	0.442
20	5.682	5.475	6.144	6.260	6.760	6.0642	1.285

14-4. ***Minting Quarters: Constructing an R Chart*** Using the same process data from Experiment 14-3, construct an R chart and determine whether the process variation is within statistical control. If it is not, identify which of the three out-of-control criteria lead to rejection of statistically stable variation.

14-5. ***Minting Quarters: Constructing an $\bar{x}$ Chart*** Using the same process data from Experiment 14-3, construct an $\bar{x}$ chart and determine whether the process mean is within statistical control. If it is not, identify which of the three out-of-control criteria lead to rejection of a statistically stable mean. Does this process need corrective action?

14-6. ***p Chart for Deaths from Infectious Diseases*** In each of 13 consecutive and recent years 100,000 children aged 0–4 years were randomly selected and the number who died from infectious diseases is recorded, with the results given below (based on data from "Trends in Infectious Diseases Mortality in the United States," by Pinner et al., *Journal of the American Medical Association,* Vol. 275, No. 3). Use STATDISK to construct a p chart. Do the results suggest a problem that should be corrected?

Number who died: 30 29 29 27 23 25 25 23 24 25 25 24 23

14-7. ***p Chart for Victims of Crime*** For each of 20 consecutive and recent years, 1000 adults were randomly selected and surveyed. Each value below is the number who were victims of violent crime (based on data from the U.S. Department of Justice, Bureau of Justice Statistics). Use STATDISK to construct a p chart. Do the data suggest a problem that should be corrected?

29 33 24 29 27 33 36 22 25 24 31 31 27 23 30 35 26 31 32 24

14-8. ***p Chart for Boston Rainfall*** Refer to the Boston rainfall amounts in Data Set 10 of Appendix B of the textbook. Delete the last value for Wednesday, so that there are 52 weeks of seven days each. For each of the 52 weeks, let the sample proportion be the proportion of days that it rained. In the first week for example, the sample proportion is 3/7 = 0.429. Use STATDISK to construct a p chart. Do the data represent a statistically stable process?

14-9. ***Polling*** When the Infopop polling organization conducts a telephone survey, a call is considered to be a defect if the respondent is unavailable or refuses to answer questions. For one particular poll about consumer preferences, 200 people are called each day, and the numbers of defects are listed below. Does the calling process require corrective action?

Defects: 92 83 85 87 98 108 96 115 121 125 112 127 109 131 130

STATDISK's Menu Configuration

File

- Open
- Close
- Save As...
- Export to Excel
- Page Setup
- Print
- Quit

Edit

- Copy
- Paste
- Help - Copy/Paste
- Sort data
- Edit column titles
- Preferences
- Printout Label

Analysis

- Probability Distributions
  - Normal Distribution
  - Student t Distribution
  - Chi-Square Distribution
  - F Distribution
  - Binomial Distribution
  - Poisson Distribution
- Binomial Probabilities
- Poisson Probabilities
- Sample Size Determination
- Confidence Intervals
- Hypothesis Testing
- Correlation and Regression
- Multiple Regression
- Multinomial Experiments
- Contingency Tables
- One-Way Analysis of Variance
- Two-Way Analysis of Variance
- Sign Tests
- Wilcoxon Tests
- Kruskal-Wallis Test
- Rank Correlation
- Runs Test
- Fisher Exact Test
- McNemar's Test
- Odds Ratio and Relative Risk
- Sensitivity and Specificity

Data

- Sample Editor
- Sample Transformations
- Sort Data
- Descriptive Statistics
- Explore Data
- Histogram
- Boxplot
- Normal Quantile Plot
- Scatterplot
- Pie Chart
- Normal Generator
- Uniform Generator
- Binomial Generator
- Poisson Generator
- Coins Generator
- Dice Generator
- Frequency Table Generator

Datasets

Help

- Statdisk Help
- Read me
- Report a bug
- What's new?
- AWL/Triola Web Site
- Check for Updates
- About Statdisk

Index